ARMY SPECIAL OPERATIONS

Change 1

August 2019
DISTRIBUTION RESTRICTION:
Approved for public release; distribution is unlimited.
This publication supersedes ADP 3-05 and ADRP 3-05, both dated 29 January 2018.
United States Government, US Army

This page intentionally left blank.

ADP 3-05, C1

Change 1
Army Doctrine Publication
No. 3-05

Headquarters
Department of the Army
Washington, DC, 26 August 2019

Army Special Operations

1. Change Army Doctrine Publication (ADP) 3-05, dated 31 July 2019, as follows:

Remove old pages:	**Insert new pages:**
v through vii (reverse blank)	v through vii (reverse blank)
1-15 and 1-16	1-15 and 1-16
2-3 and 2-4	2-3 and 2-4
4-1 and 4-2	4-1 and 4-2
5-1 and 5-2	5-1 and 5-2
8-3 through 8-6	8-3 through 8-6
Glossary-1 through Glossary-6	Glossary-1 through Glossary-6
References-1 through References-3 (reverse blank)	References-1 through References-3 (reverse blank)

2. A bar (**|**) marks changed material.

3. File this transmittal sheet in front of the publication.

DISTRIBUTION RESTRICTION: Approved for public release; distribution is unlimited.

FOREIGN DISCLOSURE RESTRICTION (FD 1): The material contained in this publication has been reviewed by the developers in coordination with the United States Army John F. Kennedy Special Warfare Center and School foreign disclosure authority. This product is releasable to students from all requesting foreign countries without restriction.

ADP 3-05, C1
26 August 2019

By Order of the Secretary of the Army:

JAMES C. MCCONVILLE
General, United States Army
Chief of Staff

Official:

KATHLEEN S. MILLER
Administrative Assistant
to the Secretary of the Army
1923804

DISTRIBUTION:
Active Army, Army National Guard, and United States Army Reserve: To be distributed in accordance with the initial distrubution number (IDN) 116092, requirements for ADP 3-05

Army Doctrine Publication No. 3-05		*ADP 3-05 Headquarters Department of the Army Washington, DC, 31 July 2019

Army Special Operations

Contents

Page

	PREFACE...iv	
	INTRODUCTION v	
	EXECUTIVE SUMMARY..ix	
Chapter 1	**OVERVIEW OF ARMY SPECIAL OPERATIONS** 1-1	
	Operational Environment... 1-1	
	Strategic Context for Army Special Operations... 1-2	
	Role 1-2	
	Core Competencies... 1-3	
	Army Special Operations Forces Characteristics.. 1-3	
	Employment Criteria 1-5	
	Support of Global Operations 1-6	
	Warfare 1-7	
	Range of Military Operations 1-8	
	Types of Military Operations 1-10	
	Functions of the Generating and Operating Force 1-11	
	Principles of Special Operations 1-11	
	Tenets of Army Special Operations... 1-12	
	Frameworks... 1-15	
	Special Operations Imperatives 1-17	
	Conclusion 1-19	
Chapter 2	**CORE ACTIVITIES** 2-1	
	Civil Affairs Operations 2-2	
	Countering Weapons of Mass Destruction 2-3	
	Counterinsurgency 2-4	
	Counterterrorism.. 2-5	
	Direct Action 2-5	
	Foreign Humanitarian Assistance 2-6	
	Foreign Internal Defense... 2-6	
	Military Information Support Operations.. 2-7	
	Preparation of the Environment... 2-8	

DISTRIBUTION RESTRICTION: Approved for public release; distribution is unlimited.

* This publication supersedes ADP 3-05 and ADRP 3-05, both dated 29 January 2018.

Contents

	Security Force Assistance	2-9
	Special Reconnaissance	2-10
	Unconventional Warfare	2-10
	Hostage Rescue and Recovery	2-11
Chapter 3	**COMMAND AND CONTROL STRUCTURES**	**3-1**
	Unity of Effort	3-1
	Theater of Operations Organizations	3-2
	Command and Control	3-4
	Army Service Component Command Structure	3-10
Chapter 4	**FIRES**	**4-1**
	Fires Warfighting Function	4-1
	Targeting Process Integration	4-1
	Special Operations Task Force Fires Considerations	4-2
Chapter 5	**INTELLIGENCE**	**5-1**
	Overview	5-1
	Special Operations Intelligence Criteria	5-1
	National-Level Intelligence Support	5-2
	Theater Intelligence	5-2
	Special Operations Intelligence Architecture	5-3
	Intelligence Processing, Exploitation, and Dissemination	5-4
Chapter 6	**SUSTAINMENT**	**6-1**
	Sustainment Brigade	6-1
	Planning	6-1
Chapter 7	**PROTECTION**	**7-1**
	Personnel Recovery	7-1
	Populace and Resources Control	7-2
	Risk Management	7-2
	Operations Security	7-3
	Army Health System	7-3
	Explosive Ordnance Disposal	7-4
	Chemical, Biological, Radiological, and Nuclear Operations	7-4
	Chemical, Biological, Radiological, and Nuclear Capabilities	7-4
Chapter 8	**EMPLOYMENT**	**8-1**
	Overview	8-1
	Military Engagement, Security Cooperation, and Deterrence	8-3
	Crisis Response and Limited Contingency Operations	8-3
	Large-Scale Combat Operations	8-4
	Summary	8-6
	GLOSSARY	**Glossary-1**
	REFERENCES	**References-1**
	INDEX	**Index-1**

Figures

Introductory Figure. ADP 3-05 logic chart vii

Executive Summary Figure. Nesting special operations with Army and joint operations x

Figure 1-1. Special operations activities 1-7

Figure 1-2. Functions: generating force, operating force, and operations 1-12

Figure 2-1. Army special operations forces core activities 2-1

Figure 3-1. United States Army Special Operations Command 3-11

Figure 3-2. 1st Special Forces Command (Airborne) 3-11

Figure 3-3. U.S. Army John F. Kennedy Special Warfare Center and School 3-13

Figure 3-4. Army Special Operations Aviation Command (Airborne) 3-14

Figure 3-5. 75th Ranger Regiment (Airborne) 3-14

Figure 5-1. Notional Army special operations intelligence flow for combatant commander daily operations 5-5

Figure 5-2. Notional Army special operations intelligence flow for large-scale combat operations 5-6

Figure 8-1. Strategic use of force, Army strategic roles, range of military operations, and objectives 8-2

Figure 8-2. Command and control of special operations forces in theater 8-5

Table

Table 3-1. Army special operations command and control echelons 3-5

Preface

ADP 3-05 provides a broad understanding of Army special operations. ADP 3-05 provides a foundation for how the Army meets the joint force commander's needs to achieve unified action by appropriately integrating Army conventional and special operations forces.

The principal audience for ADP 3-05 is all members of the profession of arms. Commanders and staffs of Army headquarters serving as joint task force or multinational headquarters should also refer to applicable joint or multinational doctrine concerning the range of military operations and joint or multinational forces. Trainers and educators throughout the Army will also use this publication. Senior Army leaders can use this publication to describe the contributions of Army special operations across the range of military operations to other senior Service leaders and senior government civilian leaders.

Commanders, staffs, and subordinates ensure their decisions and actions comply with applicable U.S., international, and, in some cases, host-nation laws and regulations. Commanders at all levels ensure their Soldiers operate in accordance with the law of war and the rules of engagement.

Army special operations are doctrinally linked to ADP 3-0, *Operations*, and JP 3-05, *Special Operations*. Army special operations forces conduct special operations in three contexts; as an integrated line of operation supporting Army unified land operations, as a separate Army line of effort supporting combatant commander campaign plans, and as the main or sole operation as directed by national command authorities.

ADP 3-05 uses joint terms where applicable. Selected joint and Army terms and definitions appear in both the glossary and the text. Terms for which ADP 3-05 is the proponent publication (the authority) are marked with an asterisk (*) in the glossary. Terms and definitions for which ADP 3-05 is the proponent publication are **boldfaced** in the text. For other definitions shown in the text, the term is *italicized* and the number of the proponent publication follows the definition.

Army special operations forces are those Active and Reserve Component Army forces designated by the Secretary of Defense that are specifically organized, trained, and equipped to conduct and support special operations (JP 3-05). These forces include Civil Affairs, Psychological Operations, Rangers, Special Forces, Special Mission Units, and Army special operations aviation forces assigned to the United States Army Special Operations Command—all supported by the 528th Sustainment Brigade (Special Operations) (Airborne).

ADP 3-05 applies to the Active Army, the Army National Guard/Army National Guard of the United States, and the United States Army Reserve unless otherwise stated.

The proponent of ADP 3-05 is the U.S. Army Special Operations Center of Excellence. The preparing agency is the U.S. Army Special Operations Center of Excellence, U.S. Army John F. Kennedy Special Warfare Center and School, Directorate of Training and Doctrine, Joint and Army Doctrine Integration Division. Send comments and recommendations on a DA Form 2028 (*Recommended Changes to Publications and Blank Forms*) to Commander, U.S. Army Special Operations Center of Excellence, U.S. Army John F. Kennedy Special Warfare Center and School, ATTN: AOJK-SWC-DTJ, 3004 Ardennes Street, Stop A, Fort Bragg, NC 28310-9610.

Introduction

ADP 3-05 provides the strategic context for employment of Army special operations forces and defines special operations. It describes the role of Army special operations formations. A *role* is the broad and enduring purpose for which the organization or branch is established (ADP 1-01). ADP 3-05 updates doctrine on Army special operations, to include incorporating the Army's operational concept of unified land operations in the context of large-scale combat operations. In addition, this publication provides information on the extant practices for special operations conducted in support of combatant commander campaign plans and globally integrated operations in support of higher campaign plans.

ADP 3-05 incorporates the *Army ethic* throughout its chapters. Army special operations are doctrinally and operationally linked to ADP 3-0. Introductory figure 1, page vii, depicts the correlation of unique aspects to Army special operations as they are overlayed against the operational construct the Army uses in ADP 3-0.

ADP 3-05—

- Rescinds ADRP 3-05, *Special Operations*.
- Revises the concept of regional mechanisms, more accurately aligning those topics with national and regional objectives.

ADP 3-05 provides the doctrinal foundation for Army special operations forces to design, plan, and conduct special operations across the range of military operations. It establishes a common frame of reference and offers intellectual tools that Army leaders use to plan, prepare for, execute, and assess special operations. By establishing a common approach and language for special operations, doctrine promotes a mutual understanding and enhances effectiveness during operations.

The doctrine in this publication is a guide for action rather than a set of fixed rules. In Army special operations, effective leaders recognize when and where doctrine, training, or even their experience no longer fits the situation, and they adapt accordingly. Subordinate special operations publications build upon the foundation provided by ADP 3-05. Other Army publications integrate and reference ADP 3-05 content where appropriate.

ADP 3-05 consolidates content from ADRP 3-05; the following is a summary of the individual chapters.

Chapter 1–Overview of Army Special Operations. This chapter defines and discusses special operations in the strategic context within which Army special operations forces expect to conduct operations, and it includes an overview of the operational environment. It addresses considerations for their employment in this environment and presents the authoritative laws and directives that govern special operations and special operations forces. It defines and describes the Army special operations forces' core competencies: special warfare and surgical strike. It presents the role and characteristics of Army special operations forces and the principles, tenets, employment criteria, and imperatives of special operations. This chapter provides an overview of traditional warfare and irregular warfare to provide the context for senior Department of Defense and other U.S. Government leaders as they consider the use of unique capabilities for global operations. It provides a short vignette to depict the unique capabilities of Army special operations Soldiers and the environments in which they conduct operations.

Chapter 2–Core Activities. This chapter narrows the broad definition of special operations down by describing the core activities that U.S. Army Special Operations Command has been directed to conduct. These core activities capture the fundamental purpose of Army special operations forces. The activities range from broad activities to very specific applications of capabilities. No single activity is ever conducted in isolation. It is the application of any combination of the activities that make up a special operation.

Introduction

Chapter 3–Command and Control Structures. This chapter describes the theater organizations that are relevant to the command and control of Army special operations forces. It provides descriptions of task organized special operations elements so they are recognizable and understood by theater assigned forces such as Army Service Component Commands. It describes command and control structures that enable the conduct of the core activities through special operations. These structures support joint unity of effort in the context of a combatant commander campaign plan and in the context of unified land operations and the Army's contributions to unity of effort. The chapter also provides an overview of the Army's special operations component command and its unit structures and functions from a generating and operating force context.

Chapter 4–Fires. This chapter briefly describes the Fires Warfighting Function before discussing how special operations forces integrate lethal and nonlethal fires into Joint and Army targeting processes. It closes by discussing fires considerations from a special operations task force perspective. The chapter also discusses the planning and execution of conventional and special operations forces during the conduct of unified land operations, particularly during large-scale combat operations.

Chapter 5–Intelligence. This chapter describes special operations intelligence criteria and how theater, national, service, and special operations intelligence capabilities are integrated. It provides examples of how intelligence information and coordination, requests, tasking, production, dissemination, resourcing, and support flows in large-scale combat operations and during the daily operations of a combatant commander campaign.

Chapter 6–Sustainment. Chapter 6 provides an overview of how U.S. Army Special Operations Command's sustainment brigade integrates its capabilities with theater Army Service component command's sustainment plans to fulfill support requirements for special operations across the range of military operations.

Chapter 7–Protection. This chapter discusses specific protection tasks that are focal points for Army special operations commanders and staffs. This focused content supplements that found in ADP 3-37, *Protection*, which applies to all Army unit commanders and staffs.

Chapter 8–Employment. This chapter provides more context and information on how Army special operations forces support achievement of national objectives as they conduct core activities across the range of military operations. This chapter also provides additional focus on support to large-scale combat operations in the context of unified land operations.

The ADP 3-05 logic chart is depicted on the following page.

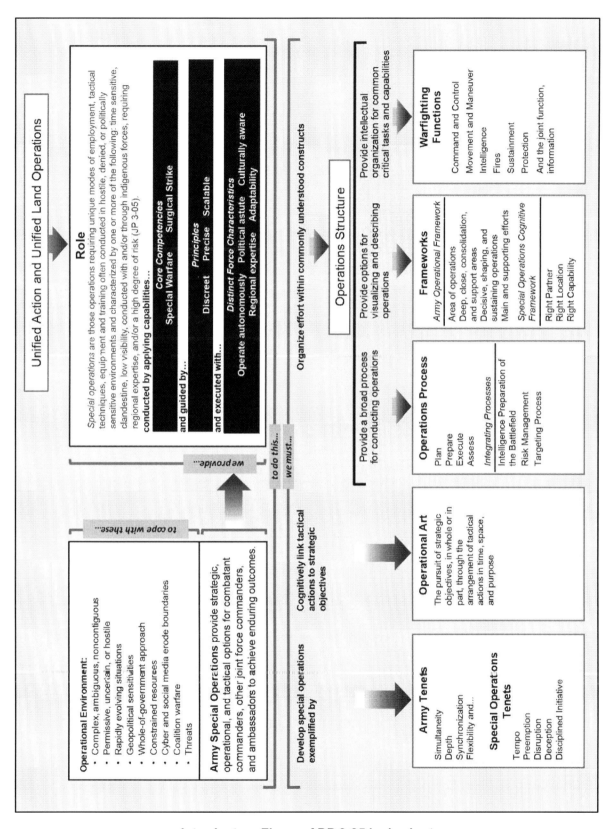

Introductory Figure. ADP 3-05 logic chart

This page intentionally left blank.

Executive Summary

This publication presents the strategic context for the conduct of Army special operations and establishes the role of Army special operations forces. A *role* is the broad and enduring purpose for which the organization or branch is established (ADP 1-01).

STRATEGIC CONTEXT

1. The strategic context for the employment of Army special operations forces consists of national policy; combatant commander, joint force commander, or ambassador requirements; the operational environment; and the threat. *Special operations* are those operations requiring unique modes of employment, tactical techniques, equipment and training often conducted in hostile, denied, or politically sensitive environments and characterized by one or more of the following: time sensitive, clandestine, low visibility, conducted with and/or through indigenous forces, requiring regional expertise, and/or a high degree of risk (JP 3-05). Special operations provide strategic, operational, and tactical options for combatant commanders, other joint force commanders, and ambassadors to achieve enduring outcomes.

FUNCTIONS OF THE GENERATING AND OPERATING FORCE

2. The U.S. Army John F. Kennedy Special Warfare Center and School, the Army's Special Operations Center of Excellence, provides specially selected, educated, and trained special operations Soldiers with regionally focused expertise. Operational commanders receive and integrate these Soldiers into their units. Their units are employed to achieve sustained engagement by executing operations in support of U.S. interests and host-nation objectives. The Army's leadership, the joint forces' leadership, and the leadership of regional partners and host nations expect sustained engagement from the Army special operations forces positioned in strategic locations around the globe. Army special operations forces provide capabilities to support the entire range of military operations conducted in support of U.S. interests and host-nation objectives. Chapter 1 discusses the lines of effort further.

ROLE

3. The leadership of the United States determines the level of required or acceptable military commitment and effort. The Army accomplishes its mission by supporting the joint force in four strategic roles: shape operational environments, prevent conflict, conduct large-scale ground combat, and consolidate gains (FM 3-0, *Operations*). The decision to conduct special operations and commit special operations forces is always validated by Chairman of the Joint Chiefs of Staff. The role of Army special operations forces is to conduct special operations. This concise role contrasts with the broad definition of special operations. JP 3-05 defines *special operations* as operations requiring unique modes of employment, tactical techniques, equipment and training often conducted in hostile, denied, or politically sensitive environments and characterized by one or more of the following: time sensitive, clandestine, low visibility, conducted with and/or through indigenous forces, requiring regional expertise, and/or a high degree of risk. This broad definition is tempered by the use of employment criteria, which facilitates decision making and prevents the unintentional expansion of tasks and responsibilities (referred to as mission creep).

4. Special operations are developed using discreet, precise, and scalable principles (executive summary figure, page x). United States Army Special Operations Command provides forces that can operate as small teams or task forces to develop and execute discreet, precise, and scalable special operations in permissive, uncertain, and hostile environments. In particular, diplomatically sensitive missions are developed and executed using the principle of being discreet. In diplomatically sensitive conditions, special operations forces create effects of a magnitude disproportionate to their small footprint. Army special operations forces support objectives that focus on deterring, preventing, or resolving joint transregional, all-domain, and multifunctional threats and conflict, as well as supporting Army operations over a multi-domain extended battlefield.

Executive Summary

CHALLENGES

- Transregional, all-domain, and multifunctional threats and conflicts
- State instability
- Proliferation of advanced technologies
- Proliferation of weapons of mass destruction
- Complex, uncertain, noncontiguous environments
- Global operations
- Whole of government approaches
- Multinational operations
- Constrained resources
- Effects created by the speed, propagation, and reach of information

To cope with these challenges United States Army Special Operations Command subordinate units support—

Unified Action and Unified Land Operations

by conducting **Special Operations**

Operations requiring unique modes of employment, tactical techniques, equipment and training often conducted in hostile, denied, or politically sensitive environments and characterized by one or more of the following: time sensitive, clandestine, low visibility, conducted with and/or through indigenous forces, requiring regional expertise, and/or a high degree of risk (JP 3-05). Special operations provide strategic, operational, and tactical options for combatant commanders, other joint force commanders, and ambassadors.

Special Operations are guided by principles
- Discreet
- Precise
- Scalable

Army special operations forces—

DESIGN AND PLAN SPECIAL OPERATIONS with the Army operations tenets of simultaneity, depth, synchronization, and flexibility and the Army special operations tenets of tempo, preemption, disruption, deception, and disciplined initiative as they use operational art to link tactical actions to strategic objectives.

Frameworks

- Provides a broad approach through the operations process—plan, prepare, execute, and assess
- Provides options for visualizing and describing operations through the Army Operational Framework—
    - Decisive, shaping, sustaining
    - Deep, close, support, consolidation
    - Main and supporting efforts
- Provides organization of common critical tasks through the Army Warfighting Functions of Mission Command, Movement and Maneuver, Intelligence, Fires, Sustainment, and Protection and the Joint Function Information
- Includes the right partner, right location, right capability cognitive framework

Army special operations forces ground the operations structure with the—

12 Special Operations Imperatives

1. Understand the operational environment
2. Recognize political implications
3. Facilitate interorganizational cooperation
4. Engage the threat discriminately
5. Anticipate long-term effects
6. Ensure legitimacy, credibility, and trust
7. Anticipate psychological effects and the impact of information
8. Operate with and through others
9. Develop multiple options
10. Ensure long-term engagement
11. Provide sufficient intelligence
12. Balance security and synchronization

Executive Summary Figure. Nesting special operations with Army and joint operations

JOINT FORCE COMMANDER REQUIREMENTS AND COMMAND AND CONTROL STRUCTURE

5. Joint force commanders (combatant or joint task force commander) identify objectives that may require the execution of special operations. Army special operations maneuver units will conduct combat operations as part of a special operations joint task force or a joint special operations task force, which may or may not be subordinate to a joint task force. A joint special operations task force may be the highest special operations echelon for limited contingency operations. Large-scale combat operations require a special operations joint task force. Army special operations units conducting military engagement, security cooperation, and deterrence operations; do so under forward-based and distributed command and control nodes; the forces and their command and control elements operate under the operational control of the commander, theater special operations command. Army special operations are scalable and the command of those operations is tailorable to the appropriate echelon for the situation as follows:

- Special operations joint task force, normally a division or corps 0-8/0-9-level command.
- Joint special operations task force, normally a group or regiment 0-6/0-7-level command.
- Special operations task force, normally a battalion 0-5-level command.
- Advanced operations base, normally a company 0-4-level command.

6. United States Special Operations Command and its subordinate commands, the United States Army Special Operations Command and the Joint Special Operations Command, advise joint force commanders on the use of their capabilities. The theater special operations commands support United States Special Operations Command core operations and activities. The theater special operations commands are trained, organized, and equipped to provide command and control of forces executing special operations.

7. Prior to designing or planning special operations and requesting Army special operations forces, commanders first determine if the objective(s) require these unique capabilities. The five criteria, discussed further in Chapter 1, used to determine this are as follows:

- It must be an appropriate special operations mission or task.
- The mission or task should support the joint force commander's campaign or operation.
- The mission or task must be operationally feasible.
- Required resources must be available.
- The outcomes must justify the risks.

PRINCIPLES

8. Army special operations forces' ability to operate in small teams in permissive, uncertain, or hostile environments allows the development and execution of special operations based on three principles. The principles of discreet, precise, and scalable special operations enable the achievement of objectives unilaterally or with or through indigenous forces and populations. These principles enable the force to conduct a wide range of missions that are often high risk and clandestine or low visibility in nature.

9. Discreet, precise, and scalable operations enhance the credibility and legitimacy of the indigenous population or host nation that special operations forces work with as follows:

- The operations are **discreet** by deliberately reducing the signature of U.S. presence or assistance.
- The operations are **precise** by using dedicated intelligence capabilities to identify, monitor, and target individuals and systems—with or without indigenous support—in an effort to reduce or eliminate collateral damage.
- The **scalable** aspect of these operations is directly associated with the way the forces are organized, trained, and equipped to carry out operations unilaterally, with minimal conventional or indigenous support. It also focuses on how the forces can execute actions that are part of a large-scale conventional operation to attain operational and strategic objectives.

IMPERATIVES

10. The following twelve imperatives are the foundation for planning and executing special operations in concert and integrated with other forces, interagency partners, and foreign organizations:
- Understand the operational environment.
- Recognize political implications.
- Facilitate interorganizational cooperation.
- Engage the threat discriminately.
- Anticipate long-term effects.
- Ensure legitimacy, credibility, and trust.
- Anticipate psychological effects and the impact of information.
- Operate with and through others.
- Develop multiple options.
- Ensure long-term engagement.
- Provide sufficient intelligence.
- Balance security and synchronization.

SUMMARY

11. ADP 3-05 updates previous doctrine to reflect the current operational environment. More importantly, this publication lays the foundation for commanders and civilian leaders to consider how Army special operations forces can be integrated to conduct discreet, precise, and scalable operations in the pursuit of national objectives.

Chapter 1

Overview of Army Special Operations

This chapter defines and discusses special operations in the strategic context within which Army special operations forces expect to conduct operations. It addresses considerations for their employment in this operational environment. It presents the role and characteristics of Army special operations forces and the principles, tenets, and imperatives of special operations.

The Army's operational concept—unified land operations—integrates the capabilities of its special operations forces with other U.S. Government efforts and activities to provide combatant commanders, other joint force commanders, and ambassadors with discreet, precise, and scalable special operations. Special operations leverage understanding of relevant actors' motivations and the underpinning of their will to achieve effects of a magnitude disproportionate to the size of the force.

OPERATIONAL ENVIRONMENT

1-1. An *operational environment* is a composite of the conditions, circumstances, and influences that affect the employment of capabilities and bear on the decisions of the commander (JP 3-0). As ADP 2-0, *Intelligence*, explains, an operational environment for a specific operation comprises more than the interacting variables that exist within a specific physical area. It also includes the interconnected global and regional influences that affect the conditions and operations with that physical area. Thus, every commander's operational environment is part of a higher commander's operational environment.

1-2. In an operational environment, the joint force operates in a complex and volatile security environment that is characterized by contested norms and persistent disorder. Transregional, all-domain, and multifunctional national security threats (such as authoritarian regimes, terrorist and criminal organizations, and non-state actors) take advantage of ungoverned and undergoverned areas and a pervasive, accessible information environment, presenting significant challenges to U.S. interests around the world. Increasingly these security threats find opportunities to exploit shared interests and combine or share capabilities. As these threats attempt to undermine legitimate governance, they foster an environment for extremism and the drive to acquire asymmetric capabilities, to include weapons of mass destruction. In addition, the threats to stability are numerous, complex, often linked, and can be aggravated by natural or man-made disasters.

1-3. Every operational environment is complex and dynamic. Information technology continues to fuel the current information revolution and makes the information environment a critical part of any operational environment that will be congested and contested in any campaign or operation. Army special operations forces are trained to consider the impact of all variables within an operational environment on any actor—allowing them to discern relevance from irrelevance and prioritize efforts on relevant actors.

1-4. Army special operations forces understand how to apply the operational variables from a strategic through a tactical perspective. They also understand the relationships between people and places, including spatial-temporal patterns, traits, and activities within the context of a geographic environment. In addition, Army special operations forces prioritize understanding the sociocultural factors that characterize the population within an operational environment. Sociocultural factors are the social, cultural, and behavioral factors characterizing the relationships and activities of the population of a specific region or operational environment (JP 2-01.3, *Joint Intelligence Preparation of the Operational Environment*, and ATP 2-01.3, *Intelligence Preparation of the Battlefield*). In order to engage with and influence civilian populations, Army special operations forces require an understanding of human factors and the ability to incorporate those factors into planning to conduct partnership activities and operations through unified action partners and indigenous populations.

Chapter 1

STRATEGIC CONTEXT FOR ARMY SPECIAL OPERATIONS

1-5. The strategic context for the employment of Army special operations forces consists of the national strategy, U.S. policy, and agreements with allies that inform combatant commander, joint force commander, or U.S. Ambassador requirements; the character of the operational environment; and the activities of adversaries or threats. *Special operations* are defined as operations requiring unique modes of employment, tactical techniques, equipment and training often conducted in hostile, denied, or politically sensitive environments and characterized by one or more of the following: time sensitive, clandestine, low visibility, conducted with and/or through indigenous forces, requiring regional expertise, and/or a high degree of risk (JP 3-05). Special operations provide strategic, operational, and tactical options to compete, challenge, and counter transregional, regional, and nation/state specific adversaries, enemies, or threats for combatant commanders, joint force commanders, and U.S. Ambassadors.

ROLE

1-6. The leadership of the United States determines the level of required or acceptable military commitment and effort. The decision to conduct special operations and commit special operations forces is always validated by the Chairman of the Joint Chiefs of Staff. The role of Army special operations forces is to conduct special operations. This concise role contrasts with the broad definition of special operations. United States Army Special Operations Command provides forces that can operate as small teams or task forces to develop and execute discreet, precise, and scalable special operations in permissive, uncertain, and hostile environments. They are trusted Army professionals of character, competence, and commitment who live by, adhere to, and uphold the moral principles of the Army Ethic and accomplish missions in the right way—ethically, effectively, and efficiently.

1-7. In particular, Army special operations forces design, plan, and conduct diplomatically sensitive missions using the principle of discreet. In diplomatically sensitive conditions, special operations forces create effects of a magnitude disproportionate to their small footprint. Special operations support objectives that focus on deterring, preventing, or resolving joint transregional, all-domain, and multifunctional threats and conflicts, as well as supporting Army operations over a multi-domain extended battlefield. Multinational and bilateral relationships with other nations' special operations forces often play a critical part for Army special operations forces executing their role. Shared doctrine, such as Standardization Agreement 2523 (North Atlantic Treaty Organization), *Allied Joint Doctrine for Special Operations*/Allied Joint Publication-3.5, *Allied Joint Doctrine for Special Operations*, facilitates interoperability, integration, and interdependence between these forces.

1-8. As part of global, transregional, and regional strategies, special operations include a range of coordinated, integrated, and synchronized activities conducted with unified action partners. Through sustained engagement with these operational partners, Army special operations forces foster information sharing, enhanced interoperability, and the collaborative execution of missions—all of which facilitate joint operational planning and execution.

1-9. Army special operations forces are postured to support Army strategic roles (shaping, preventing, stability, and the consolidation of gains) in order to prevail in all environments. They are prepared to disrupt or eliminate threats unilaterally, with partners or friendly indigenous forces, or as a component of a joint force in order to shape the deep fight and defeat forces in the close fight. During operations to consolidate gains, Army special operations forces are integral to assisting the transition of civil activities, supporting host-nation sovereignty, and setting conditions to prevent further conflict and stabilize the security environment.

1-10. Designing campaigns and joint operations for the optimal relationship between Army special operations forces, Army conventional forces, joint forces, indigenous security forces and institutions, and resistance partners can prevent future conflict or mitigate the frequency and duration of a potential future crisis. Achieving the optimal force composition and designing interdependence of Army special operations, conventional forces, joint forces, and indigenous security assets directly affect the success of all military operations. The operational environment, specific threats, and the scale of the effort lend themselves to a flexible template for blended or integrated operations that are either special operations forces-specific (only), special operations forces-centric (primarily), and/or conventional forces-centric. Interagency involvement is likely required in all circumstances. Such blended operations are more successful when they are deliberately planned to have interdependence between this broad array of capabilities. Success is facilitated when the executing force is comprised of units habitually aligned with special operations forces by region, have routinely trained together, and perhaps have an advisory cadre to augment special operations forces capabilities.

CORE COMPETENCIES

1-11. A *core competency* is an essential and enduring capability that a branch or an organization provides to Army operations (ADP 1-01). A core competency is not a task; it is a capability stated in general terms. The Army special operations forces' core competencies explain to both internal and external audiences what the force contributes to the Nation's security, to joint force commanders, and to Army commanders. The core competencies are not categories of operations nor are they meant to be attributed to one unit over another. The core competencies are not used to bin core activities. The core competencies provide overarching categories for the wide range of capabilities that Army special operations forces provide. They are enduring— whereas subordinate capabilities may change over time to meet the demands of an ever-changing global security environment. Subordinate Army special operations forces doctrine does not discuss the core competencies; it focuses on their operations and may include the branch's core competencies if they have been identified. The Army special operations forces' core competencies are special warfare and surgical strike.

1-12. *Special warfare* is **the execution of capabilities that involve a combination of lethal and nonlethal actions taken by a specially trained and educated force that has a deep understanding of cultures and foreign language, proficiency in small-unit tactics, and the ability to build and fight alongside indigenous combat formations in permissive, uncertain, or hostile environments (ADP 3-05).** Special warfare captures the unique capabilities that special operations forces have in regard to the human aspects of military operations. These capabilities are integrated into every special operation, whether those operations are conducted unilaterally or as part of a regional or global campaign. Special warfare capabilities facilitate leveraging indigenous forces, providing understanding of relevant actors, building and maintaining partnerships, and creating access.

1-13. *Surgical strike* is **the execution of capabilities in a precise manner that employ special operations forces in hostile, denied, or politically sensitive environments to seize, destroy, capture, exploit, recover or damage designated targets, or influence threats (ADP 3-05).** Executed unilaterally or collaboratively, surgical strike capabilities are applied to extend operational reach and influence by engaging global targets using the principles of discreet, precise, and a scalable. Surgical strike capabilities should not be confused with tasks or core activities such as direct action. Surgical strike capabilities are applied to shape the operational environment or influence a threat target audience in support of larger strategic interests.

ARMY SPECIAL OPERATIONS FORCES CHARACTERISTICS

1-14. A *characteristic* is a feature or quality that marks an organization or function as distinctive or is representative of that organization or function (ADP 1-01). Army special operations forces are characterized in a manner that underscores their utility and uniqueness. Understanding these characteristics allows combatant commanders, other joint force commanders, and ambassadors to develop a greater appreciation of how the force applies its capabilities through special operations to achieve a broad range of national objectives. The principles of special operations—discreet, precise, and scalable—guide the planning of special operations, and the characteristics of special operations forces allow them to design, plan, and execute those operations.

1-15. Army special operations forces are trained to operate independently in small teams. From the group/brigade perspective, the unit of action is the team. However, execution of special operations missions and tasks require companies and teams to be able to shift their perspective to consider that an individual special operations Soldier can be a unit of action. The ability to operate autonomously is a characteristic that applies to all Army special operations forces, and it is a key attribute that allows the force to be employed in a package that produces the smallest possible footprint.

1-16. The characteristics of being politically astute and culturally aware contribute to the force's ability to build and sustain relationships with indigenous partners. These attributes are specifically developed so that Army special operations forces can meet the requirements demanded by political and cultural sensitivities often associated with the introduction of U.S. forces. These characteristics facilitate the ability to design, plan, and conduct operations and activities that deliberately create effects in the environment, enable friendly forces and regional partners, and enable an enduring friendly force presence. These operations and activities provide a foundation for Army shaping operations and decisive operations, particularly when the environment degrades and requires large-scale combat operations.

1-17. Army special operations units are aligned regionally. However, simply aligning a unit to a region is not enough to fulfill combatant commander and ambassador requirements. Regional expertise is a characteristic developed over time that increases exponentially from individual to team level. Regionally aligning units establishes a foundation for developing regional expertise, allowing Soldiers to progress in rank and responsibility within a unit and continue to develop their expertise over time. Regional expertise facilitates time-sensitive operational requirements by decreasing the amount of time required to understand the operational environment. Combining regional expertise with the other Army special operations forces characteristics provides combatant commanders with a force that can deploy rapidly and provide ground truth to the combatant command, as well as advise U.S. Ambassadors in rapid developing situations. Providing early advice and military options facilitates critical decision making by the ambassador and the country team, often de-escalating the situation and preventing excessive loss of life.

Army Special Operations Assessment and Selection

SGT Johnson is an infantry team leader who after 3 years of service, which included overseas combat tours, decides to re-enlist. Having been exposed to Army special operations forces from Civil Affairs, Psychological Operations, and Special Forces units during training and deployment, the sergeant decides to attend Psychological Operations Assessment and Selection. SGT Johnson is selected, and attends the qualification course. Upon graduating, the sergeant is assigned to the 7th Psychological Operations Battalion.

Fast forward 6 months—the sergeant is deployed to a country in Africa on a predeployment site survey. SGT Johnson will be the noncommissioned officer in charge of the upcoming mission. As part of the site survey, the sergeant is responsible for briefing the U.S. Ambassador of the country, the Defense Attaché, the regional security officer, the host-nation Deputy of Minister of Defense and the host-nation Minister of Information. SGT Johnson coordinates with the U.S. Embassy staff, spending extra time with the finance office to ensure the transfer of mission funds is coordinated and to confirm U.S. Embassy processes for contracts with host-nation vendors. The sergeant visits the site where they will conduct operations from, meets the host-nation mission partners, and coordinates sustainment and training requirements.

On the final day of the site survey, SGT Johnson boards a commercial aircraft and settles in. As the plane lifts off, the sergeant reflects. Just 2 years ago, he would not have imagined briefing anyone higher than a company commander or conducting a leader's reconnaissance for anything but a patrol base. SGT Johnson considers the impact of his decision to attend the Psychological Operations Qualification Course and how it prepared him for transition to this unit and for these types of missions. The sergeant makes a mental note to visit the schoolhouse and thank the instructors and the detachment and section sergeants for the premission training.

SGT Johnson settles back in his seat and remembers the apprehension he felt a week ago when departing on that first mission as a Psychological Operations Soldier. The sergeant relaxes, recognizing that the training fully prepared him to help lead the mission, and he starts planning the training for the team deployment.

1-18. Adaptability complements the other characteristics. Adaptability includes adapting to unknown physical environments, adapting to diverse cultures, adapting to perceptions, and adapting to cognitive processes. Adaptability allows a commander to place a team in an unfamiliar region with unfamiliar cultures and achieve desired results. The individuals' and team's collective adaptability allows them to rapidly apply the foundational training that enabled their other characteristics (such as understanding culture versus understanding a culture) and be successful in any area of operations across geographic areas of responsibilities.

1-19. No single characteristic is more important than another. The Army special operations assessment and selection process ensures that candidates possess the capacity to develop each characteristic and apply them

Overview of Army Special Operations

together. It is the application of these characteristics that make the Army special operations force distinct from other forces and of great utility to combatant commanders, joint force commanders, and ambassadors. The depth of these characteristics increases over time, with training, education, and experience. These characteristics enable Army special operations forces to design, plan, and conduct discreet, precise, and scalable special operations for permissive, uncertain, and hostile environments.

EMPLOYMENT CRITERIA

1-20. Joint force commanders—a combatant commander or a joint task force commander—identify objectives that may require special operations force capabilities. Army special operations maneuver units will conduct combat operations as part of a special operations joint task force or a joint special operations task force, which may or may not be subordinate to a joint task force. A joint special operations task force may be the highest special operations echelon for limited contingency operations, but large-scale combat operations require a special operations joint task force.

1-21. Army special operations units conducting military engagement, security cooperation, and deterrence operations, do so under forward-based and distributed command and control nodes; the forces and their command and control elements operate under the operational control of the Commander, theater special operations command. Army special operations units with the mission to close with and destroy an enemy are the U.S. Army Special Forces Groups, Special Mission Units (a generic term to represent an organization composed of operations and support personnel that is task-organized to perform highly classified activities [JP 3-05]), and the 75th Ranger Regiment. Psychological Operations and Civil Affairs units provide unique capabilities in support of both Army special operations maneuver units, conventional forces, and U.S. Ambassadors.

1-22. Prior to designing or planning special operations and requesting Army special operations forces in support of a combatant commander's campaign plan or a contingency plan, commanders first determine if the objective(s) require the unique capabilities of Army special operations forces. As trusted Army professionals and stewards of the Army Profession, commanders conduct a thorough analysis to ensure that highly trained, competent special operations forces are employed ethically, effectively, and efficiently to accomplish the most important and highest payoff missions in the right way.

1-23. Commanders consider five basic operational mission criteria when determining whether Army special operations forces are the appropriate force for their operational requirement:

- **The mission must be an appropriate special operations force mission or task.** Army special operations forces should be used to create effects that require their unique skills and capabilities. They should not be used as a substitute for other forces. If the effects do not require those skills and capabilities, then commanders should not request or assign special operations forces.
- **The mission or tasks should support the joint force commander's campaign or operation plan.** If the mission does not support the campaign or major operation plan, other courses of actions should be considered.
- **The mission or tasks must be operationally feasible.** The Army's special operations force is not structured for attrition or traditional force-on-force tactics and should not be assigned missions beyond their capabilities. Commanders and their staffs must consider the vulnerability of the units to larger, more heavily armed, or more mobile forces, and particularly in a hostile environment.
- **Required resources must be available to execute and support the mission.** Some missions require support from other forces for success. Support involves aiding, protecting, complementing, and sustaining employed Army special operations forces. Support can include airlift, intelligence, communications, influence activities, medical, logistics, space capabilities, weather effects, and numerous other types of support. Although a target may be vulnerable to Army special operations unit capabilities, deficiencies in supportability may affect the likelihood for success or may entirely invalidate the feasibility of employing the force.

Chapter 1

- **The expected outcome of the mission must justify the risks.** Special operations forces are of high value and are limited in numbers. While Army special operations forces constitute over half of the Department of Defense special operations force, they are only five percent of the total Regular Army strength. Commanders must make sure the benefits of successful mission execution are measurable, create significant opportunities for the joint force commander, and are in balance with the risks inherent in the mission. Commanders should recognize the high value and limited resources of Army special operations forces. *Risk management*, the process to identify, assess, and control risks and make decisions that balance risk cost with mission benefits (JP 3-0), considers not only the potential loss of units and equipment but also the potential adverse effects on U.S. diplomatic and political interests, if the mission fails or is publicly exposed. Although special operations may present the potential for a proportionally greater effect on a campaign or operation, there may be occasions in which conditions restrain special operations to a marginal net effect, while assuming too much risk through the loss of personnel and materiel.

SUPPORT OF GLOBAL OPERATIONS

1-24. The U.S. military is engaged in one of the most challenging periods in its history. Army special operations forces are—and will be—continuously engaged around the globe. Army special operations units execute activities specified in Section 167 of Title 10, United States Code; other similar activities as may be specified by the President or the Secretary of Defense; and special operations forces functions and responsibilities described in other directives. Army special operations forces provide combatant commanders precise, lethal and nonlethal capabilities. The special operations activities are shown in progression from legislation to doctrine in figure 1-1, page 1-7.

1-25. Army special operations support global operations by providing forces trained and equipped to support the Commander, United States Special Operations Command and other combatant commanders' efforts. Army special operations commanders employ their forces to execute precise, discreet, scalable missions and activities to prevent and deter conflict or to prevail in war. These missions and activities are executed in the context of national objectives related to assessment, shaping, active deterrence, influence, disruption, and threat neutralization.

1-26. Special operations are a key component of globally integrated operations. Special operations forces provide critical information and produce intelligence products that are instrumental in supporting joint transregional, all-domain, multi-functional threats and conflicts and Army operations over a multi-domain extended battlefield. Such information and intelligence assist commanders in developing appropriate lines of effort and lines of operations to destroy threat networks, shape conditions, deter adversaries, and influence relevant actors. Special operations can be executed unilaterally, in conjunction with indigenous forces, in conjunction with joint conventional forces, or in conjunction with joint conventional and indigenous forces.

1-27. The United States Army Special Operations Command and the United States Army John F. Kennedy Special Warfare Center and School, provide Army conventional force commanders, joint force commanders, and U.S. Ambassadors with forces that can execute discreet, precise, and scalable special operations in permissive, uncertain, or hostile environments. This organized, trained, and equipped force contributes to unified effort at every echelon. The force can operate in denied or sensitive areas to collect, monitor, and verify information of strategic and operational significance, using a variety of tactics and techniques to meet the requirements of discreet operations. The activities listed in figure 1-1, page 1-7, generate information that special operations forces can provide directly to a joint force commander or ambassador and integrate into intelligence processes. The resulting products are disseminated to military and other governmental departments and agencies, as well as indigenous military, police, or other personnel to facilitate execution of their missions.

Overview of Army Special Operations

United States Code Title 10, Section 167 Unified Combatant Command for Special Operations Forces	Department of Defense Directive 5100.01 Functions of the Department of Defense and its Major Components	Joint Publication 3-05 Special Operations
• Civil Affairs • Counterterrorism • Direct action • Foreign internal defense • Humanitarian assistance • Military information support operations • Strategic reconnaissance • Such other activities as may be specified by the President or the Secretary of Defense • Theater search and rescue • Unconventional warfare	Subject to this authority, direction and control of the Secretary of Defense, Commander, U.S. Special Operations Command is responsible for and has the authority necessary to conduct, in addition to those specified, all affairs of such command relating to special operations activities, including— • Civil Affairs operations • Counterproliferation of weapons of mass destruction • Counterproliferation operations • Counterinsurgency • Direct action • Foreign internal defense • Information operations • Military information support operations • Security force assistance • Special reconnaissance • Unconventional warfare	• Civil Affairs operations • Countering weapons of mass destruction • Counterinsurgency • Counterterrorism • Direct action • Foreign humanitarian assistance • Foreign internal defense • Hostage rescue and recovery • Military information support operations • Preparation of the environment • Security force assistance • Special reconnaissance • Unconventional warfare

Figure 1-1. Special operations activities

WARFARE

1-28. The strategic environment is extremely fluid, with continually changing coalitions; alliances; partnerships; and transregional, all-domain, multifunctional national security threats. Global operations are the implementation of a national security strategy (way) and are provided the necessary resources (means) to achieve desired outcomes (ends) in the strategic environment. Global operations apply strategic uses of force including assurance, deterrence, and coercion. War requires the strategic use of military force—compellence. War is a violent clash of wills predicated on the strategic interests (perceived needs and aspirations) of a state or non-state actor. The two forms of warfare recognized by the U.S. military are traditional and irregular, and the U.S. military recognizes that the most effective application of the two forms is applying them in combination. Traditional warfare is the predominate form of warfare used when opposing actors have symmetrical capabilities. Irregular warfare has a focus on gaining and maintaining legitimacy, credibility, and influence with a relevant population.

1-29. *Irregular warfare* is defined as a violent struggle between state and non-state actors for legitimacy and influence over the relevant population(s) (JP 1). Irregular warfare is typically the predominant form of warfare used when the actors have asymmetric capabilities, making the population a critical factor and a potential center of gravity. Another critical factor is will. As previously mentioned, will and interest are connected. However, will can tip the probability of success in the favor of a weaker actor over a stronger actor. Army special operations forces use their understanding of relevant actors' motivations and the underpinning of relevant actors' will when designing and planning special operations. This understanding places Army special operations forces in a position to provide key contributions to the joint force no matter how the combination of traditional and irregular warfare is applied.

1-30. Irregular warfare may be between states, between state and non-state actors, or between non-state actors with no state involvement. Nation-states and/or non-state actors may conduct irregular warfare as an element of competition between adversaries, as a component of international armed conflict, as an application of contingency response, or as a distinct armed conflict. Irregular warfare mission areas include, but are not limited to, foreign internal defense, unconventional warfare, counterterrorism, counterinsurgency, and stabilization. Key enabling activities may include, but are not limited to, countering threat networks, counter threat finance, civil-military operations, Civil Affairs operations, security force assistance, intelligence operations, cyberspace operations, military information support operations, operations conducted in and through the information environment, and support to law enforcement. As depicted in figure 1-1, page 1-7, there is an association of these mission areas and enabling activities with the capabilities of special operations forces.

1-31. Most joint and Army forces can be employed in any of the irregular warfare contexts described above. Army special operations forces have an irregular warfare focus, in that they are specifically designed and optimized for irregular warfare contexts, purposes, and mission areas, but can also be applied in a traditional warfare context. A combination of irregular warfare mission areas have contributed to the military defeat of an enemy and subsequent stabilization of the region or state in armed conflict (major combat operations, crisis response, and limited contingency operations). Some examples include World War II, Operation DESERT STORM, Operation IRAQI FREEDOM, and Operation ENDURING FREEDOM-PHILIPPINES. Army special operations forces, therefore, will continue to contribute to the Army's success in large-scale combat by integrating irregular warfare mission areas as supporting efforts, enabling traditional warfare activities. Army conventional forces—designed and optimized for traditional warfare contexts, purposes, and missions—can apply those capabilities in an irregular warfare context.

1-32. The combination of Army conventional and special operations forces is the norm in most Army operations. In any operation, Army special operations forces pair their strategic to tactical understanding of human factors, populations, human geography, and other relevant aspects of an operational environment with the capability to conduct low-visibility operations with and through foreign forces. This technique is the foundation for conducting the entire range of activities and operations that fall under the umbrella of irregular warfare. The unique capabilities of the varied special operations units are applied together to conduct these and other population-centric activities and operations across the entire competition continuum (cooperation, competition below armed conflict, and armed conflict) unilaterally or with unified action partners.

RANGE OF MILITARY OPERATIONS

1-33. The United States employs Army special operations capabilities in support of U.S. national security objectives in a variety of operations. These operations vary in size, purpose, risk, and conflict-intensity within a range of military operations. These operations extend from military engagement, security cooperation, and deterrence to crisis response and limited contingency operations and, if necessary, large-scale combat operations. To generalize, U.S. national security objectives focus on protecting U.S. interests. The Chairman, Joint Chiefs of Staff and his staff analyze these objectives and communicate them as military national objectives. Combatant commanders, their component commands, subordinate Army special operations headquarters, and the executing special operations unit refine these objectives for their echelon. The national objectives supported by Army special operations units are typically related to assessment, shaping, active deterrence, influence, disruption, and threat neutralization.

1-34. Use of Army special operations capabilities in military engagement, security cooperation, and deterrence activities support combatant commander campaign objectives to effect the operational environment; maintain U.S. influence, access, and interoperability with and to designated actors; and maintain or achieve stability in a region. Many of the missions associated with limited contingencies, such as logistics support, foreign humanitarian assistance, and defense support of civil authorities, do not require combat. However, some such operations can rapidly escalate to combat operations and require a significant effort to protect U.S. forces while accomplishing the mission. Individual, major operations often contribute

Overview of Army Special Operations

to a larger, long-term effort—for example, those that are part of global operations. The nature of the strategic environment is such that special operations forces are often engaged in several types of joint operations simultaneously. Army special operations forces are specifically organized, manned, trained, and equipped to execute specified core activities. These activities are conducted across the range of military operations as supporting or primary activities, either singly or in combination. These core activities help provide a bridge across the range of operations, helping to create stable security environments and—when the environment degrades due to crisis or armed conflict—providing a foundation from which positions of relative advantage are obtainable.

Note: Chapter 2, *Core Activities*, provides additional details.

MILITARY ENGAGEMENT, SECURITY COOPERATION, AND DETERRENCE

1-35. Military engagement, security cooperation, and deterrence are usually associated with permissive environments. Army special operations forces execute these activities to establish, shape, maintain, and refine relationships with other nations and foreign and domestic civil authorities. They develop information and intelligence, which contributes to shared situational, regional, and global awareness and ultimately a shared understanding. Shared understanding enables activities that shape operational environments and help maintain U.S. influence, access, or interoperability with and to actors or regions.

1-36. *Military engagement* is the routine contact and interaction between individuals or elements of the Armed Forces of the United States and those of another nation's armed forces, or foreign and domestic civilian authorities or agencies to build trust and confidence, share information, coordinate mutual activities, and maintain influence (JP 3-0). Army special operations forces' military engagement activities influence and assist a nation and its institutions in order to further U.S. objectives in the region. These efforts support a broad range of indigenous capabilities, capacity, and conditions that enable—

- Host-nation cooperation as a U.S., regional, and global partner.
- Development of a population and a national resilience.
- An increase in the capacity and capability of host-nation security forces and their supporting institutions.
- Host-nation forces to, unilaterally or in partnership with Army special operations forces, defeat subversion, lawlessness, insurgency, terrorism, and other threats to its security within their sovereign borders.

1-37. *Security cooperation* is defined as all Department of Defense interactions with foreign security establishments to build security relationships that promote specific United States security interests, develop allied and partner nation military and security capabilities for self-defense and multinational operations, and provide United States forces with peacetime and contingency access to allied and partner nations (JP 3-20). The participation of Army special operations forces in security cooperation missions fosters relationships with host-nation forces and key leaders through sustained contact. Security cooperation and partner activities are key elements of global and combatant commander campaigns.

1-38. Deterrence helps prevent adversary action through the presentation of a credible capability and willingness of counteraction. Effective deterrence requires a combatant commander campaign plan, which coordinates communication efforts; integrates military engagement and security cooperation activities; and synchronizes with U.S. Government foreign assistance efforts to create favorable conditions for U.S. forces and partner nations. Army special operations forces' capabilities contribute to all deterrence efforts, ranging from show of force, joint exercises, antiterrorism, support to counterdrug efforts, flexible deterrent options, and flexible response options. A critical aspect of deterrence is sustained presence in the operating environment. The Army special operations forces' capacity, in regards to sustained presence, directly relates to the force's ability to conduct scalable operations.

1-39. Because Army special operation forces routinely execute activities in small teams, they provide combatant commanders, other joint force commanders, and ambassadors an efficient way to create a sustained presence. In addition, Army special operations forces provide key capabilities for *nonproliferation*—actions to prevent the acquisition of weapons of mass destruction by dissuading or impeding access to, or distribution of, sensitive technologies, material, and expertise (JP 3-40).

Chapter 1

U.S. Special Operations Command directs U.S. Army Special Operations Command to conduct *counterproliferation*—those actions taken to reduce the risks posed by extant weapons of mass destruction to the United States, allies, and partners (JP 3-40) as a core activity.

CRISIS RESPONSE AND LIMITED CONTINGENCY OPERATIONS

1-40. JP 3-0 describes crisis response and limited contingency operations as typically focused in scope and scale and conducted to achieve a very specific strategic or operational-level objective in an operational area. Included are operations to ensure the safety of American citizens and U.S. interests, while maintaining and improving the ability of the United States to operate with multinational partners in deterring the hostile ambitions of potential aggressors. Army special operations forces serve honorably to protect American citizens and demonstrate courage by doing what is right despite risk, uncertainty, and fear. They can lead or support either type of operation, whether conducted as a standalone operation to resolve a crisis such as Operation PROVIDE COMFORT or as an element of a larger, more complex operation such as Operation SUPPORT DEMOCRACY. Typical crisis response and limited contingency operations include noncombatant evacuation operations, peace operations, foreign humanitarian assistance, recovery operations, strikes, raids, homeland defense, and defense support to civilian authorities.

1-41. The capability to execute assigned core activities and the ability to achieve expertise in the required skills and enabling activities—such as advanced special operations techniques, identity intelligence, tactical combat casualty care, and rotary-wing fires, infiltration, and exfiltration—allow Army special operations forces executing crisis response and limited contingency operations to seamlessly transition to support large-scale combat operations.

LARGE-SCALE COMBAT OPERATIONS

1-42. When required to achieve national strategic objectives or to protect national interests, the U.S. national leadership may decide to conduct a major operation involving large-scale combat, placing the United States in a wartime state. In such cases, the general goal is to prevail against the enemy as quickly as possible, to conclude hostilities, and to establish conditions favorable to the United States, and its multinational partners. Large-scale combat operations are usually a blend of traditional and irregular warfare activities. The expertise of Army special operations forces described in the paragraphs on crisis response and limited contingency operations apply equally to large-scale combat operations. The principles of discreet, precise, and scalable facilitate the integration of special operations into unified land operations as the Army conducts large-scale combat operations. Scalability allows the Army special operations force to increase the scale of special operations as the (joint) area of operations transitions to large-scale combat operations.

1-43. Consolidating gains made prior to large-scale combat operations positions the force to conduct discreet and precise special operations—
- In denied areas to leverage indigenous populations and other human networks.
- To open denied areas.
- To thwart threat anti-access efforts.
- To facilitate and execute deep operations for joint task force component commanders.
- To provide sensors, combat information, and intelligence from beyond the fire support coordination line.
- To conduct combat identification to inform engagement decisions.

1-44. Special operations conducted beyond fire support coordination lines and outside of the land force area of operations place Army special operations forces at risk of friendly fires. Therefore, detailed coordination is conducted to establish restrictive fire support coordination measures as well as to determine where to place a fire support coordination line.

TYPES OF MILITARY OPERATIONS

1-45. Army doctrine addresses decisive action as described in ADP 3-0. Army commanders at all echelons may combine different types of operations simultaneously and sequentially to accomplish missions. For each mission, the joint force commander and Army component commander determine the emphasis Army forces place on each type of operation. Missions in any environment require Army special operations forces to be prepared to conduct any combination of offensive, defensive, stability, or defense support of civil authorities tasks.

FUNCTIONS OF THE GENERATING AND OPERATING FORCE

1-46. Special operations provide strategic, operational, and tactical options for combatant commanders, other joint force commanders, and ambassadors to achieve enduring outcomes. Special operations complement the Army's ability to provide a force that is postured to shape and influence through global special operations and develop a global special operations forces network that is prepared to conduct combat operations as part of the joint fight. The global special operations forces network is a synchronized network of people and technology (United States, allies, and partner nations) designed to support commanders through inter-operable capabilities that enable special operations (JP 3-05).

1-47. Three functions guide the development and employment of special operations:
- **Generating a Force Through Purposeful Investment.** The United States Army Special Operations Center of Excellence, the generating force, provides specially selected, educated, and trained special operations Soldiers with regionally focused expertise. Generating a force with regional expertise, cross-cultural communication skills, and targeted linguistic capabilities is instrumental in the institutional training of all special operations forces prior to their assignment to the operational force.
- **Operating a Force With Sustained Engagement.** Operational commanders integrate these Soldiers into the operating force and employ their units to achieve sustained engagement and to execute operations in support of U.S. interests and host-nation objectives. The leadership of the Army, joint forces, regional partners, and host nations expect sustained engagement from Army special operations forces positioned in strategic locations around the globe.
- **Executing Operations in Support of U.S. Interests and Host-Nation Objectives.** Army special operations forces provide capabilities to support the entire range of military operations conducted in support of U.S. interests and host-nation objectives.

1-48. Figure 1-2, page 1-12, shows the construct that U.S. Army Special Operations Command uses to establish its doctrine and focus the command on its evaluation and development of the doctrine, organization, training, materiel, leadership and education, personnel, and facilities domains require to effectively support the Army and the joint force.

PRINCIPLES OF SPECIAL OPERATIONS

1-49. Army special operations forces' ability to operate in small teams in permissive, uncertain, or hostile environments allows the development and execution of special operations based on core principles. The core principles of discreet, precise, and scalable special operations enable the achievement of objectives unilaterally, or with or through indigenous forces and populations. These principles enable the force to conduct a wide range of missions that often have high risk, are clandestine, or require a posture of low visibility and help characterize special operations. Discreet, precise, and scalable operations provide combatant commanders and ambassadors a flexible application of military capabilities in politically sensitive and culturally complex environments.

1-50. These operations enhance the credibility and legitimacy of the indigenous population or host nation that special operations forces work with as follows:
- The operations are **discreet** by deliberately reducing the signature of U.S. presence or assistance.
- The operations are **precise** by using dedicated intelligence capabilities to identify, monitor, and target individuals and systems, with or without indigenous support, while eliminating collateral damage.
- The **scalable** aspect of these operations is directly associated with the way the forces are organized, trained, and equipped to carry out operations unilaterally with minimal conventional or indigenous support. It also focuses on how the forces can execute actions that are part of a large-scale conventional operation to attain operational and strategic objectives.

1-51. The principles used to develop special operations are applied across the range of military operations. The principles enable the force to engage directly with partner nations as the U.S. Government and those partners contend with transregional, all-domain, multifunctional competition. When used in conjunction with other government efforts, special operations increase the possibility of maintaining states of cooperation and competition and avoid escalation to armed conflict. However, when armed conflict is unavoidable, the principles of special operations allow Army special operations forces to execute operations to facilitate an end to armed conflict by increasing the lethality of U.S. and partner forces, while simultaneously mitigating risks to the joint and multinational force.

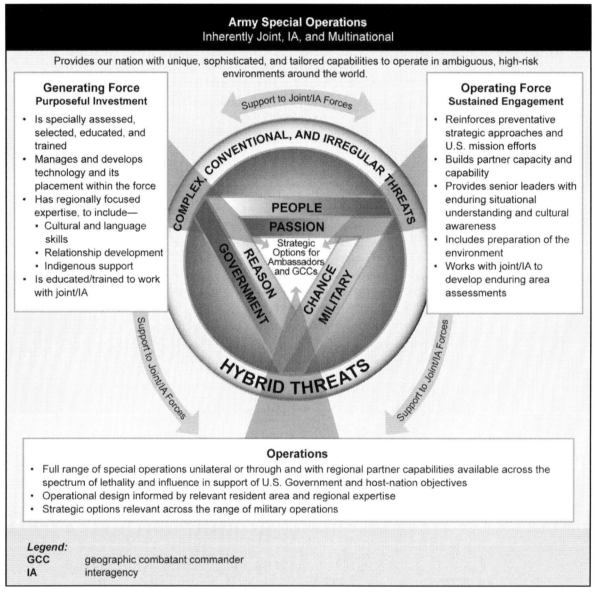

Figure 1-2. Functions: generating force, operating force, and operations

TENETS OF ARMY SPECIAL OPERATIONS

1-52. To execute discreet, precise, and scalable special operations, commanders and staffs design and plan special operations with the Army operations tenets of simultaneity, depth, synchronization, and flexibility. In addition, they plan operations using the Army special operations tenets of tempo, preemption, disruption, deception, and disciplined initiative as they use operational art to link tactical actions to strategic objectives.

TEMPO

1-53. Tempo is critical to the ability of the force to conduct special operations that require a direct approach. A rapid tempo of execution with respect to the enemy allows special operations forces to mass their combat power at the critical place and time, execute a task, withdraw before the enemy can react, and then repeat that execution process until the mission is accomplished. A rapid tempo offsets small numbers and limited firepower by not allowing the enemy the time to bring its main strength to bear on the committed special operations force. At the same time, a rapid tempo provides a degree of security through speed, by offsetting a higher degree of risk than might otherwise be accepted.

PREEMPTION

1-54. Special operations can preempt an adversary by neutralizing the adversary's capabilities before the fight, either directly or in support of conventional forces, through—

- Foreign internal defense efforts to build indigenous defense, influence, and intelligence capabilities.
- Unconventional warfare activities that enable a resistance movement.
- Security force assistance activities that build the capability and capacity of regional partners, facilitate regional security conditions, and provide the foundation for coalition operations.
- Military information support operations that influence relevant actors, to include enemy, adversary, friendly, and neutral organizations' leadership, armed forces, and populations, to create conditions that preclude the use of an enemy's/adversary's capabilities.
- Civil Affairs operations that engage and leverage the civil component, mitigate threats to civil society, and increase a host-nation's legitimacy, capability, and capacity to address population grievances.
- Counterproliferation activities to slow or inhibit development of a capability.
- Direct action missions against the enemy's critical, strategic or operational strengths.
- Counterterrorist operations to neutralize the use of third parties by a state sponsor and/or to neutralize a threat categorized as a terrorist organization.
- Lessons learned from previously executed preparation of the environment activities and special operations intelligence products to provide combatant commanders, joint task force commanders, and ambassadors with first-hand information, validated intelligence on population, and other relevant actor perspectives, intentions, strengths, vulnerabilities, susceptibilities, opportunities, issues, goals, and influence.

DISRUPTION

1-55. Army special operations forces can unilaterally disrupt a threat. They can also conduct special operations with interagency and intergovernmental partners, partner nation forces, friendly indigenous forces, or as a component of a joint force to disrupt a threat. These operations and missions are conducted throughout the range of military operations and during any phase of a large-scale combat operation. Army special operations disrupt the adversaries' ability to use supporting physical and human networks; shadow governments; economic, financial, and intelligence infrastructures; and the ability to make timely, informed decisions.

1-56. The objective that purposeful disruption supports may be to induce conditions that influence a relevant actor to behave in a manner that is favorable for the joint force. Examples of these behaviors include making tactical or strategic decisions that make the relevant actor vulnerable to joint force lethal activities and taking actions that isolate the relevant actor from his center of gravity. Operations and missions intended to influence a relevant actor's behavior may require longer execution times than other disruption activities and may start prior to operations for armed conflict.

1-57. Whether conducted unilaterally or with joint, interagency, intergovernmental, indigenous, and multinational partners, Army special operations forces can achieve disruptive effects by conducting—

- Special reconnaissance to collect or verify information of strategic or operational significance and facilitate U.S. or partner capabilities to exploit that information and disrupt the relevant actor.

Chapter 1

- Direct actions against strategically and operationally significant targets and adversaries' centers of gravity.
- Military information support operations to influence specified target audiences; disrupt alliances; segregate the adversary from the population; incite dissension, desertion, malingering, and sabotage within the adversaries armed force; foster resistance movements; and foster the resiliency of friendly populations and other relevant actors.
- Population and resources control to provide security for the populace, deny personnel and material to the threat, mobilize population and material resources, and detect and reduce the effectiveness of threats.
- Human geographic analysis and network engagement to support, influence, or compel populations, governments, and other institutions to expose malign influence; counter coercion and subversion; and impose costs through conventional and unconventional activities.
- Unconventional warfare activities to enable specified elements of a resistance movement to execute disruption tasks unilaterally, bilaterally, with a multinational force, or with U.S. forces in an advisory capacity. Special Forces units possess the capabilities to organize and direct large indigenous forces that cause the enemy to spread its forces thin.
- Foreign internal defense by assisting a host nation to organize and direct large indigenous forces that create a military shield behind which other governmental departments and agencies can operate to remove the causes of insurgency.

DECEPTION

1-58. Army special operations forces can provide operational-level commanders multiple means to attack an adversary's will. Army special operations forces can create perceptions that there are too many forces for the adversary to counter effectively. With no safe areas and adversary forces subject to attack anywhere at any time, an adversary's will can be significantly weakened. Army special operations forces may conduct or support tactical deception, military deception, or information-based operations that will cause the enemy to redeploy or dislocate in a manner favorable to friendly forces.

DISCIPLINED INITIATIVE

1-59. Army special operations unit leaders foster a professional organizational climate in which all live by and uphold the Army Ethic. The Army Ethic establishes the foundation for accomplishing the mission in the right way and in an environment of mutual trust—one that encourages critical thinking, ethical reasoning, adherence to and upholding the Army Ethic, freedom of action, and disciplined initiative in subordinates. This environment provides a force with the ability to make independent, time-critical decisions using all available information and guidance presented in higher headquarters commander's intent. Successful missions result from subordinate leaders at all echelons understanding the commander's intent, assessing and accepting risks, and exercising disciplined initiative to accomplish missions. Army special operations forces at any echelon describe how their tasks support objectives—up to and including national objectives. Special operations unit commanders use the mission command approach to command and control in order to—

- Establish the conditions that strengthen mutual trust and cohesive teamwork.
- Enable and empower agile and adaptive subordinate leaders to exercise disciplined initiative within the commander's intent.
- Ensure shared understanding of how the execution of their tasks, missions, and operations nest within each higher echelon's plan and objectives.

1-60. Although Army special operations personnel must be included in centralized planning at the combatant commander and subordinate joint task force commander levels, successful special operations require decentralized planning. Army special operations forces command, plan, and conduct operations through the mission command approach by conducting decentralized planning and execution of special operations down to the team level. Mission command enables special operations units down to the team level to conduct planned operations that higher echelons account for. Colloquially, this is called bottoms up planning. When the executing special operations unit is charged with conducting a critical mission that higher echelons depend on, particularly when these missions are unilateral, commanders provide their mission statement and

Overview of Army Special Operations

intent and let the operators plan the mission. This is an inherent element of the mission command approach, which allows those commanders with the best situational understanding to plan and—once a mission is approved for execution—make rapid decisions without waiting for higher echelon commanders to assess the situation and issue orders.

1-61. When there are multiple executing special operations units, special operations task forces and higher echelons must ensure the selected courses of action are complimentary to each other and supportable. A *special operations task force* is a scalable unit, normally of battalion size, in charge of the special operations element, organized around the nucleus of special operations forces and support elements (JP 3-05). Independent judgment and effective coordination by leaders at every echelon are vital to successful special operations.

FRAMEWORKS

1-62. All Army forces undergo standardized training and education in troop leading procedures, the military decision making process, the mission command approach, the warfighting functions, and the operations process. Army special operations forces receive additional training and education in the joint processes that parallel Army processes. Army special operations forces use this training and education and structurally apply it to special operations conducted in support of Army, joint, or other Service commanders, and U.S. Ambassadors. The organization of common critical tasks through the Army warfighting functions of mission command, movement and maneuver, intelligence, fires, sustainment, and protection and the joint function of information—combined with the operations process of plan, prepare, execute, and assess—result in a unique operational approach. In addition, the operational approach is developed by applying planning considerations for special operations that require working with or through a foreign partner. Army special operations planners use a cognitive framework to determine how the right partner, at the right location, with the right capability enable the efforts in achieving a combatant commander's campaign plan objectives and achieving the objectives in an ambassador's country strategy. When large-scale combat operations are required, Army special operations planners leverage previous missions, particularly those that involved foreign partners, and visualize and describe operations through the Army operational framework—decisive, shaping, and sustaining operations; deep, close, support, and consolidation areas; and main and supporting efforts.

COGNITIVE FRAMEWORK

1-63. The right partner includes identifying, assessing, and selecting a partner force for training, advisement, or assistance that can enhance U.S. national security. Army special operations forces work with or through indigenous partners who are best able to advance U.S. strategic objectives and can directly or indirectly support the combatant commander's campaign plans and the ambassadors' integrated country strategies. These critical partnerships are based on trust and a shared desire to achieve mutually beneficial outcomes. The right partner possesses the necessary level of motivation, access, and placement required to facilitate shaping and deterrence operations. A partner does not have to be a state actor. Army special operations forces routinely forge partnerships with resistance movements, autonomous tribes, and a variety of other non-state actors in order to enable access and influence throughout the operational area.

1-64. The right locations include identifying geographic locations with strategic or operational significance or suitability. Army special operations forces operate in locations that have a strategic significance and support national objectives in an effort to deter conflict and shape the operational environment. Access to key populations allows the forces to exert influence and, therefore, alter the conditions in the operational environment. Since they are culturally adept and politically astute, they can maneuver the intricate terrain of population-centric activities and are able to capitalize on emerging opportunities. When special operations are conducted in response to emerging crises, the location is dictated by the situation itself. In such cases, they select optimal locations to stage and deploy forces in order to maximize effectiveness and to mitigate risks.

1-65. The right capability includes enhancing a partner's capability to augment the security of the United States and its allies. When working with a partner, Army special operations forces continuously assess the partner's requirements to identify critical capability gaps. Practical solutions and appropriate mechanisms are applied to mitigate validated capability gaps. The partner's capabilities are enhanced to a level that is sustainable and ensures success of the U.S. and partner collaborative effort. Programs designed to enhance a partner's capability are tailored to address the partner's specific requirements and not exceed the partner's ability to sustain it. A well-designed program ensures that the partner possesses sufficient capability to support U.S. efforts, can use the capability appropriately, and can sustain that capability after U.S. forces depart.

ARMY OPERATIONAL FRAMEWORK AREAS

1-66. The *deep area* is where the commander sets conditions for future success in close combat. (ADP 3-0). As part of the joint force, a special operations joint task force will be assigned an area of operation that may include joint land component commander or corps deep areas. Special operations in these areas include leveraging indigenous personnel, direct action, intelligence operations, target identification, and air space control; coordinating joint fires; supporting resistance movements; creating conditions to enable conventional force maneuvers; and conducting other activities to shape the deep fight. Special operations conducted in deep areas require constant communication between all joint force staffs (land, air, maritime, and special operations) down to brigade combat team level. This communication is facilitated by the exchange of liaison officers who are empowered to coordinate on behalf of their commanders.

1-67. The *close area* is the portion of the commander's area of operations where the majority of subordinate maneuver forces conduct close combat. (ADP 3-0). While historically there has been doctrinal emphasis on coordination of activities in deep areas, the same emphasis should be considered for activities in close areas. Close areas are often the most fluid portion of the joint operational area as friendly and enemy forces seek to conduct decisive operations. Civilian populations within these areas add further complexity to the situation. Army special operations forces, particularly Civil Affairs and Psychological Operations units, may be attached to conventional force commanders in a number of command and control relationships.

Note: JP 1 provides additional information on command and control relationships such as operational and tactical control.

1-68. Special Forces, Ranger, special operations aviation and special operations sustainment units, who are planning operations or located in a conventional force commander's area of operation, must coordinate their operations and activities with that commander. Army special operations forces that directly support Army conventional force commanders in the close area (brigade combat team, battalion, company) exercise due diligence in ensuring that their integration into the unit includes checks on interoperability. In addition, this diligence ensures that their operations support the higher echelon operational approach for conventional and special operations force interdependence. This diligence is a shared responsibility of the receiving unit and the attached unit. For special operations units not directly supporting conventional force commanders in the close area, liaison exchange is so important that it can be viewed as a requirement to ensure coordination of fire control measures, movement control measures, maneuver, obstacle locations, and a shared common operational picture.

1-69. A *support area* is the portion of the commander's area of operations that is designated to facilitate the positioning, employment, and protection of base sustainment assets required to sustain, enable, and control operations (ADP 3-0). Regardless of whether special operations units are based with or separate from conventional forces, they will likely fall within the sustainment area of an Army commander. Special operations sustainment elements facilitate planning the sustainment requirements during the operations process at the theater level. However, the fluidity of large-scale combat operations requires these sustainment personnel to integrate with the sustainment elements of the support area commander to properly coordinate support.

Note: Chapter 3 provides an overview on special operations sustainment elements and Army special operations task-organized elements. Chapter 6 provides more information on special operations sustainment.

1-70. A *consolidation area* is the portion of the land commander's area of operations that may be designated to facilitate freedom of action, consolidate gains through decisive action, and set conditions to transition the area of operations to follow on forces or other legitimate authorities (ADP 3-0). In the consolidation area, special operations are integral to assisting the transition of civil activities, supporting host-nation sovereignty, and setting conditions to prevent further conflict and stabilize the security environment. Drawing on their ability to work closely with foreign security forces, Army special operations forces may prove uniquely suited to identifying and neutralizing bypassed enemy forces attempting to organize continued or new resistance to joint operations. To consolidate gains and stabilize the environment, additional Army special operations forces may be introduced to the joint operations area. Civil Affairs and Psychological Operations forces may augment existing conventional forces or work directly with other legitimate authorities to consolidate gains. These forces are specially trained, equipped, and organized to support and conduct stability activities, Civil Affairs operations, and civil-military operations, as well as to influence populations. Not only do these activities contribute to stabilization, but by influencing local populations, remaining enemy forces, and supporting other relevant actors, Army special operations forces ensure minimal interference with continued friendly operations and enable freedom of action in the close area. Like operations elsewhere, Army special operations forces must continue to coordinate their efforts with the commander assigned responsibility for the consolidation area. This coordination facilitates unity of effort by keeping everyone informed of the joint forces' efforts.

SPECIAL OPERATIONS IMPERATIVES

1-71. Whereas the special operations principles characterize successful special operations, the imperatives discussed below prescribe key operational requirements. The imperatives are fundamental rules that guide how the force approaches the design, planning, and conduct of missions. These rules guide how special operations Soldiers think about their tasks, missions, and operations. These imperatives can provide the basis for establishing foundational activities for the force as it plans and executes special operations in concert with other forces, interagency partners, and foreign organizations:

- **Understand the Operational Environment.** Special operations cannot shape the operational environment without first gaining a clear understanding of the theater of operations, to include civilian influence and enemy and friendly capabilities. Special operations forces achieve objectives by understanding the political, military, economic, social, information, infrastructure, physical environment, and time variables within the specific operational environment and by developing plans to act within the realities of those operational environments. Army special operations forces must identify the friendly and hostile decision makers, their objectives and strategies, and the ways in which they interact. The conditions of conflict can change, and they must anticipate these changes in the operational environment and exploit fleeting opportunities.

- **Recognize Political Implications.** Many special operations are conducted to advance critical political objectives. Army special operations forces understand that their actions can have international consequences. Whether conducting operations independently or in coordination with partners, they must consider the political effects of their actions. They must anticipate ambiguous operational environments in which military factors are not the only concern. Special operations frequently create conditions for nonmilitary activities to occur within indigenous populations and for host-nation civil institutions to achieve U.S. and host-nation objectives. Consequently, the advancement of political objectives may take precedence over the achievement of military priorities.

- **Facilitate Interorganizational Cooperation.** Most special operations occur in an interorganizational environment in which the U.S. Government departments and agencies, multinational partners, and international organizations are supporting common national and mutual security objectives as part of a country team effort. Army special operations forces must actively and continuously coordinate their activities with all relevant parties—U.S. and foreign military and nonmilitary organizations—to ensure efficient use of all available resources and maintain unity of effort.

- **Engage the Threat Discriminately.** Special operations missions often have sensitive political implications. Therefore, commanders must carefully select when, where, and how to employ the force. Special operations are executed with precision and accuracy to minimize collateral effects and in a concealed or clandestine manner including through the actions of indigenous partners so that only the effects are detectable.

- **Anticipate Long-Term Effects.** Army special operations forces must consider the broader political, economic, informational, cultural, demographic, and military effects when faced with dilemmas because the solutions will have broad, far-reaching effects. These forces must accept legal and political constraints to avoid strategic failure while achieving tactical success. They must not jeopardize the success of national and combatant command long-term objectives with a desire for immediate or short-term effects. Policies, plans, and operations must be consistent with the national and theater of operations priorities and objectives they support. Inconsistency can lead to a loss of legitimacy and credibility at the national and international levels.

- **Ensure Legitimacy, Credibility, and Trust.** Significant legal and policy considerations apply to many special operations activities. Legitimacy and mutual trust are the most crucial factors in developing and maintaining internal and international support. The United States cannot sustain its assistance to a foreign power without legitimacy. Commanders, staffs, and subordinates foster legitimacy, credibility, and trust through decisions and actions that adhere to the moral principles of the Army Ethic and comply with applicable U.S., international, and, in some cases, host-nation laws and regulations. Commanders at all levels are responsible for the conduct of their Soldiers who must operate in accordance with the law of war and the rules of engagement. However, the concept of legitimacy is broader than the strict adherence to law. The concept also includes the moral-ethical and political legitimacy of a government or resistance organization. The people of the nation and the international community determine legitimacy of the host-nation government or cause, based on their collective acceptance of its right and authority to govern and exercise power. The perceived legitimacy of its cause and the ethical application of its power, based upon international rules of law, are strong factors in determining political legitimacy. However, it is the decisions and actions of trusted Army professionals who demonstrate their character, competence, and commitment by adhering to and upholding the Army Ethic who create and maintain international and host-nation trust in the legitimacy of Army special operations. Without shared understanding, mutual trust, and a credible host-nation government, legitimacy is not possible. Without legitimacy, credibility, and trust, Army special operations forces will not receive the support of the indigenous elements that is essential to success.

- **Anticipate Psychological Effects and the Impact of Information.** Special operations forces have always anticipated the psychological effects of their operations and attempted to conduct those operations in a manner that helps control those effects. The information joint function is a recognition of this on a macro level—it accounts for the effects of information on the operational environment as well as on specified relevant actors. A deep understanding of the operational environment and the relevant actors in it is required to anticipate psychological effects. Understanding the underlying reasons that drive these effects allows the force to understand why the effects may occur. This provides a foundation to understand potential behavior and responses to friendly force activities. By understanding causes and anticipating effects, the force can determine which actions are advantageous or disadvantageous to friendly force objectives. Understanding the psychological effects of special operations activities on the environment is part of leveraging information and implementing the joint function—information.

- **Operate With and Through Others.** Army special operations forces seek opportunities to create partners and leverage the partners' capabilities. During multinational operations, there may be opportunities to advise, train, and assist military and paramilitary forces. The supported non-U.S. forces then serve as force multipliers in the pursuit of mutual security objectives with minimum U.S. visibility, risk, and cost. Army special operations forces operate with and through indigenous government and civil society leaders to shape the operational environment. The long-term self-sufficiency of the foreign forces and entities requires them to assume primary authority and accept responsibility for the success or failure of the mission. All U.S. efforts must reinforce and enhance the capacity, effectiveness, legitimacy, credibility, and trust of the supported foreign government or group.

- **Develop Multiple Options.** Army special operations forces maintain their operational flexibility by developing a broad range of options and contingency plans. They must be able to shift from one option to another before and during mission execution, or apply two or more simultaneously, to provide flexible, national and regional options.

- **Ensure Long-Term Engagement.** Army special operations forces recognize the need for persistence, patience, and continuity of effort. They demonstrate continuity of effort when dealing with political, economic, informational, and military programs. Special operations policy, strategy, and programs must, therefore, be durable, consistent, and sustainable.

- **Provide Sufficient Intelligence.** Success for special operations missions and tasks dictates that uncertainty associated with the threat and other aspects of the operational environment must be minimized through the application of intelligence operations and procedures. Because of the needed detailed intelligence, Army special operations forces typically must also access theater of operations and national intelligence systems to address intelligence shortfalls and to ensure that timely, relevant, accurate, and predictive intelligence is on hand. Human intelligence is often the primary intelligence discipline for satisfying priority intelligence requirements, whether from overt or controlled sources. The key to effective intelligence support is for special operations forces to fully leverage the entire intelligence support system and architecture. Army special operations units also provide information and intelligence through area assessments, special reconnaissance, and postoperation debriefs.
- **Balance Security and Synchronization.** Security concerns often dominate special operations, but over-compartmentalization can exclude key special operations forces and conventional forces personnel from the planning cycle. Special operations unit commanders must resolve these conflicting demands on mission planning and execution. Insufficient security may compromise a mission; conversely, excessive security may also jeopardize a mission.

CONCLUSION

1-72. The challenges in the security environment are numerous and include—
- Transregional, all-domain, and multifunctional threats and conflicts.
- Unstable states.
- Proliferation of advanced technologies and weapons of mass destruction.
- Complex, uncertain, and noncontiguous operational environments.
- Constrained resources.
- Effects created by the speed, propagation, and reach of information.

1-73. Army special operations units support unified action and unified land operations by conducting special operations, thereby contributing to the joint force and Army efforts to overcome these challenges. Special operations are guided by the principles of discreet, precise, and scalable. Army special operations forces design and plan special operations with the Army operations tenets of simultaneity, depth, synchronization, and flexibility and the Army special operations tenets of tempo, pre-emption, disruption, deception, and disciplined initiative as they use operational art to link tactical actions to strategic objectives. They use an operations structure that allows a broad approach through the Army operations process; provides options for visualizing and describing operations through the Army Operational Framework; organizes common critical tasks through the Army Warfighting Functions and the Joint Function—Information; and integrates the right partner, right location, right capabilities framework. Special operations and the related principles, tenets, and operations structure are grounded by the twelve Army special operations forces imperatives.

This page intentionally left blank.

Chapter 2
Core Activities

Army special operations forces possess unique capabilities to support U.S. Special Operations Command's roles, missions, and functions as directed by Congress in Title 10, United States Code, Section 164, *Commanders of Combatant Commands: Assignment; Powers and Duties,* and Section 167, *Unified Combatant Command for Special Operations Forces*. Army special operations forces design, plan, conduct, and support special operations throughout the range of military operations. These missions are normally joint or interagency in nature. Army special operations forces can conduct these missions unilaterally, with allied forces, as a coalition force, or with indigenous assets. Mission priorities vary from one theater of operations to another. The missions are dynamic because they are directly affected by politico-military considerations. A change in the National Security Strategy or policy may add, delete, or radically alter the nature of an Army special operations forces unit's mission.

The President, the Secretary of Defense, or a joint force commander may task an Army special operations element to perform missions for which it is the best suited among available forces, or perhaps the only force available.

2-1. Special operations forces conduct core activities using unique capabilities under conditions in which other forces are not trained or equipped to execute. Army special operations forces are specifically organized, manned, trained, and equipped to accomplish twelve U.S. Special Operations Command directed activities. In addition to these twelve core activities, select Army special operations units are also trained and equipped to conduct hostage rescue and recovery. Figure 2-1 provides a list of all Army special operations forces core activities.

Army Special Operations Forces Core Activities

- Civil Affairs operations
- Countering weapons of mass destruction
- Counterinsurgency
- Counterterrorism
- Direct action
- Foreign humanitarian assistance
- Foreign internal defense
- Military information support operations
- Preparation of the environment
- Security force assistance
- Special reconnaissance
- Unconventional warfare
- Hostage rescue and recovery (select Army special operations units only)

Figure 2-1. Army special operations forces core activities

Chapter 2

2-2. The Army special operations forces' core activities are derived from Title 10, United States Code. The Secretary of Defense, in DODD 5100.01, *Functions of the Department of Defense and its Major Components*, assigned these activities to Commander, U.S. Special Operations Command. Commander, U.S. Special Operations Command, in U.S. Special Operations Command Directive 10-1, *Terms of Reference—Roles, Missions, and Functions of Component Commands*, directed these activities to Commander, U.S. Army Special Operations Command.

2-3. In addition, Commander, U.S. Special Operations Command designated U.S. Army Special Operations Command as the lead special operations component for the following skills and activities:

- Airborne operations (military free-fall and static line).
- Advanced special operations techniques.
- Civil Affairs operations.
- Counterinsurgency.
- Foreign internal defense (ground and rotary-wing aircraft).
- Identity intelligence operations.
- Joint special operations medical training.
- Military information support operations.
- Mountain operations.
- Multipurpose canine.
- Personnel recovery, nonconventional assisted recovery, and unconventional assisted recovery.
- Preparation of the environment.
- Rotary-wing fires, infiltration, and exfiltration techniques.
- Security force assistance (ground and rotary-wing aircraft).
- Sensitive site exploitation.
- Sniper operations.
- Special operations urban combat.
- Unconventional warfare.

CIVIL AFFAIRS OPERATIONS

2-4. *Civil Affairs operations* are actions planned, coordinated, executed, and assessed to enhance awareness of, and manage the interaction with, the civil component of the operational environment; identify and mitigate underlying causes of instability within civil society; and/or involve the application of functional specialty skills normally the responsibility of civil government (JP 3-57). Civil Affairs operations are performed by assigned Civil Affairs forces. *Civil Affairs* are designated Active Component and Reserve Component forces and units organized, trained, and equipped specifically to conduct Civil Affairs operations and to support civil-military operations (JP 3-57). Military commanders must consider not only military forces but also the entirety of the operational environment in which they operate. This operational environment includes a populace that may be supportive, neutral, or antagonistic to the presence of military forces, both friendly and opposing. A supportive populace can provide resources and information to facilitate friendly operations. Civil Affairs operations ensure that commanders are advised on civil considerations, to include fulfillment of legal and moral obligations to the populace (in conjunction with the commander's legal advisor), which can impact foreign policy objectives. Commanders and staffs use the principle of precise operations to ensure that noncombatants are treated with dignity and respect, minimizing harm to them and protecting their personal property, and to establish trust with the host-nation populace. A hostile populace threatens the operations of deployed friendly forces, can undermine mission legitimacy, and can often undermine public support at home for the nation's policy objectives. *Civil-military operations* are activities of a commander performed by designated military forces that establish, maintain, influence, or exploit relations between military forces and indigenous populations and institutions by directly supporting the achievement of objectives relating to the reestablishment or maintenance of stability within a region or host nation (JP 3-57).

2-5. Civil Affairs units conduct Civil Affairs Operations to support the commander's civil-military operations efforts. These forces are the commander's primary asset to purposefully engage nonmilitary organizations, institutions, and populations. Civil Affairs units establish, maintain, influence, or exploit relations between military forces and civil authorities (government and nongovernment) and the civilian populace in a friendly, neutral, or hostile area of operations to facilitate military operations, shape the environment, prevent future conflict, maintain operational tempo, preserve combat power, and to consolidate gains. Civil Affairs units may assist or perform activities and functions that are normally the responsibility of local government. Civil Affairs operations may occur before or during military operations or postconflict and are of particular importance in consolidation areas. They may also occur, if directed, in the absence of other military operations.

2-6. Civil Affairs units engage and influence the civil component of the operational environment through Civil Affairs activities, military government operations, and Civil Affairs–supported activities in order to provide civil-considerations expertise through the planning and execution of these operations, and enable civil-military operations. The forces are organized, trained, and equipped specifically to plan and execute Civil Affairs operations across the range of military operations, engaging many different civil entities (indigenous populations and institutions, unified action partners, and interagency partners). In limited instances, they also involve the application of functional specialty skills, by U.S. Army Reserve Civil Affairs forces, in areas normally the responsibility of civil government, which enhance the conduct of operations. The units are organized, equipped, and trained to carry out missions that specifically include the conduct of Civil Affairs operations.

Note: FM 3-57, *Civil Affairs Operations*, provides Army commanders with the information necessary for the integration of Civil Affairs capabilities in support of unified land operations. It provides the doctrinal basis for the conduct of Civil Affairs operations in support of civil-military operations.

COUNTERING WEAPONS OF MASS DESTRUCTION

2-7. U.S. Special Operations Command is designated lead combatant command and coordinating authority for all Department of Defense countering weapons of mass destruction planning efforts in coordination with other Combatant Commands, Military Services, and as directed by other appropriate U.S. Government departments and agencies (DODD 2060.02, *DOD Countering Weapons of Mass Destruction [WMD] Policy and the Unified Command Plan*). This role is separate and distinct from those combatant command and special operations duties/responsibilities defined in Title 10, Sections 164 and 167. *Weapons of mass destruction* are chemical, biological, radiological, or nuclear weapons capable of a high order of destruction or causing mass casualties, and excluding the means of transporting or propelling the weapon where such means is a separable and divisible part from the weapon (JP 3-40).

2-8. *Countering weapons of mass destruction* is defined as efforts against actors of concern to curtail the conceptualization, development, possession, proliferation, use, and effects of weapons of mass destruction, related expertise, materials, technologies, and means of delivery (JP 3-40). The Department of Defense supports interagency partners' nonproliferation activities, which includes actions to prevent the acquisition of weapons of mass destruction by dissuading or impeding access to, or distribution of, sensitive technologies, materials, and expertise. *Counterproliferation* are those actions to reduce the risks posed by extant weapons of mass destruction to the United States, allies, and partners (JP 3-40). Weapons of mass destruction pathway defeat is an operational approach to conducting deliberate actions against actors of concern and their networks to dissuade, destroy, or otherwise complicate the conceptualization, development, possession, and proliferation of weapons of mass destruction, related expertise, materials, technologies, and means of delivery. Weapons of mass destruction defeat are operations to seize, neutralize, or destroy existing weapons of mass destruction and means of deliver to deny possession and prevent transfer or employment. The priority objectives outlined in the *Department of Defense Strategy for Countering Weapons of Mass Destruction* are to—

- Reduce incentives to pursue, possess, and employ weapons of mass destruction.
- Increase barriers to the acquisition, proliferation, and use of weapons of mass destruction.
- Manage weapons of mass destruction risks emanating from hostile, fragile, or failed states and safe havens.
- Deny the effects of current and emerging weapons of mass destruction threats through layered, integrated defenses.

Chapter 2

2-9. The pursuit of weapons of mass destruction and its potential use by actors of concern poses a threat to U.S. national security and peace and stability around the world. Army special operations forces' capabilities include—

- Providing expertise, material, and teams to supported combatant commands to locate, identify, tag, and track weapons of mass destruction, as required.
- Conducting direct action operations in limited access areas, as required.
- Building partnership capacity for conducting counterproliferation activities.
- Conducting special reconnaissance.
- Conducting counterterrorism operations.
- Employing military information support operations in conjunction with national and regional level influence efforts to dissuade adversary reliance or use of weapons of mass destruction.
- Conducting counter threat finance activities.
- Providing other specialized countering weapons of mass destruction capabilities.

COUNTERINSURGENCY

2-10. Resistance to legitimate government authority in the form of insurgency has the potential to be a large and growing element of the security challenges that the United States faces in the 21st century. *Counterinsurgency* is defined as comprehensive civilian and military efforts designed to simultaneously defeat and contain insurgency and address its root causes (JP 3-24). Counterinsurgency operations can be the primary effort or a line of operation in support of a larger operation. Successful operations are population-focused because of the importance of building support for the government and its programs. Likewise, the population is a center of gravity for an insurgency and targeted as part of an integrated counterinsurgency effort. Army special operations forces are a principal U.S. military contribution to these operations, leveraging previous engagements and preparation of the environment activities with the host nation to provide a host-nation approved approach. The force operate discreetly in local communities, directly communicating with target audiences and influencing their behaviors. They can also conduct complex counterterrorist operations in support of the overarching counterinsurgency operation.

2-11. To be successful, Army special operations unit commanders must understand the underlying causes of violence, the core grievances that the insurgency is fighting for, and the structure of the insurgency movement. Army special operations forces are particularly valuable because of their specialized capabilities:

- Civil Affairs operations.
- Cultural expertise.
- Deception (both tactical and military deception).
- Intelligence operations.
- Military information support operations.
- Region-specific knowledge.
- Expertise in the human factors of resistance movements.

2-12. Army special operations units execute counterinsurgency operations and related activities, and its missions and tasks have two primary lines of effort. They must assist the host-nation forces to defeat or neutralize the insurgent militarily. This assistance allows the host-nation government to start or resume functioning in once-contested or insurgent-controlled areas. Army special operations forces support the overall effort by conducting security force assistance, military information support operations, training, intelligence, and tactical support. This provides an environment in which the host-nation government can win the trust and support of its people and become self-sustaining. Both aspects of the mission are of equal importance and must be conducted at the same time. However, if the host nation does not fully address the underlying causes of violence and the grievances of the insurgency, short-term tactical gains will be temporary.

COUNTERTERRORISM

2-13. *Terrorism* is the unlawful use of violence or threat of violence, often motivated by religious, political, or other ideological beliefs, to instill fear and coerce governments or societies in pursuit of goals that are usually political (JP 3-07.2). *Counterterrorism* is defined as activities and operations taken to neutralize terrorists and their organizations and networks in order to render them incapable of using violence to instill fear and coerce governments or societies to achieve their goals (JP 3-26). Army special operations forces possess the capability to conduct these operations in environments that may be denied to conventional forces because of political or threat conditions. Army special operations forces are trusted Army professionals who serve with honor and apply their military expertise in order to ensure they are continuously and consistently making the right decisions and taking the right actions to deter, disrupt, neutralize, and defeat terrorists.

2-14. Host-nation responsibilities, Department of Justice and Department of State lead agency authorities, legal and political restrictions, and appropriate Department of Defense directives dictate Army special operations forces involvement in counterterrorism. Special operations are used to conduct offensive measures within the Department of Defense's overall efforts.

2-15. Army special operations units conduct counterterrorism missions as special operations by overt, covert, clandestine, or low-visibility means. Activities include—

- Intelligence operations to collect, exploit, and report information on terrorist organizations, personnel, assets, and activities.
- Network and infrastructure attacks to execute preemptive strikes against terrorist organizations. The objective is to destroy, disorganize, disrupt, or disarm terrorist organizations before they can strike targets of U.S. national interests and interests of allied nations.
- Hostage rescue recovery or sensitive materiel recovery from terrorist control. These activities require capabilities not normally found in conventional military units. Ensuring the safety of the hostages and preventing destruction of the sensitive materiel are essential mission requirements.
- Nonlethal activities to defeat the ideologies or motivations that spawn terrorism by nonlethal means. These activities could include military information support operations, Civil Affairs operations, unconventional warfare activities, security force assistance, and foreign internal defense activities.

DIRECT ACTION

2-16. *Direct action* is defined as short-duration strikes and other small-scale offensive actions conducted as a special operation in hostile, denied, or diplomatically sensitive environments and which employ specialized military capabilities to seize, destroy, capture, exploit, recover, or damage designated targets (JP 3-05). Direct action differs from conventional offensive actions in the level of physical and political risk, operational techniques, and the degree of discriminate and precise use of force to achieve specific objectives. In the conduct of these operations, special operations units may—

- Employ raid, ambush, or direct assault tactics (including close quarters battle).
- Emplace mines and other munitions.
- Conduct standoff attacks by fire from air, ground, or maritime platforms.
- Provide terminal guidance for precision-guided munitions.
- Conduct independent sabotage.
- Conduct anti-ship operations.

2-17. Normally limited in scope and duration, direct action operations usually incorporate an immediate withdrawal from the planned objective area. These operations can provide specific, well-defined, and often time-sensitive results of critical, strategic and operational significance.

2-18. Army special operations forces may conduct direct action operations independently or as part of larger conventional or unconventional operation. Although normally considered close-combat-type operations, direct action operations also include sniping and other standoff attacks by fire delivered or directed by special operations forces. Standoff attacks are preferred when the target can be damaged or destroyed without close combat. The force employs close-combat tactics and techniques when the mission requires precise or discriminate use of force or recovery or capture of personnel or materiel.

2-19. Direct action missions may also support locating, recovering, and restoring to friendly control selected persons or materiel that are isolated and threatened in sensitive, denied, or contested areas. These missions usually result from situations that involve political sensitivity or military criticality of the personnel or materiel being recovered from remote or hostile environments. These situations may arise from a political change, combat action, chance happening, or mechanical mishap. Army special operations forces can use close quarters battle in direct action operations. ***Close quarters battle*** is **sustained combative tactics, techniques, and procedures employed by small, highly trained special operations forces using special purpose weapons, munitions, and demolitions to recover specified personnel, equipment, or material (ADP 3-05)**. Direct action operations differ from the personnel recovery method of combat search and rescue by the use of—

- Dedicated ground combat elements.
- Unconventional techniques.
- Accurate, precise, and predictive intelligence.
- Indigenous assistance.

2-20. Direct action operations may be unilateral or multinational, but the operations are still short-duration, discreet actions. Direct action operations are conducted within a special operations command and control structure to achieve the supported commander's objectives. Unlike unconventional warfare operations, they do not involve the support of an indigenous chain of command to achieve objectives of mutual interest.

FOREIGN HUMANITARIAN ASSISTANCE

2-21. *Foreign humanitarian assistance* consists of Department of Defense activities conducted outside the United States and its territories to directly relieve or reduce human suffering, disease, hunger, or privation (JP 3-29). While U.S. military forces are not the primary U.S. Government means of providing humanitarian assistance, the assistance they are tasked to provide is designed to supplement or complement the efforts of the civil authorities or agencies that have the primary responsibility for providing that assistance.

2-22. Special operations forces can deploy rapidly with long-range communications equipment and operate in the austere and chaotic environments typically associated with disaster-related humanitarian assistance efforts. Perhaps the most important capabilities found within the Army special operations force for foreign humanitarian assistance are their geographic orientation, cultural knowledge, language capabilities, and their ability to work with multiethnic indigenous populations and international relief organizations to provide initial and ongoing assessments. Civil Affairs units establish a civil-military operations center to enable the commander and staff to coordinate, collaborate, and synchronize stabilization efforts in disaster areas. Army special operations units can provide temporary support, such as airspace control for landing zones, communications nodes, security, and advance force assessments, to facilitate the deployment of conventional forces and designated humanitarian assistance organizations until the host nation or another organization can provide that support.

FOREIGN INTERNAL DEFENSE

2-23. *Foreign internal defense* is participation by civilian agencies and military forces of a government or international organizations in any of the programs and activities undertaken by a host nation government to free and protect its society from subversion, lawlessness, insurgency, terrorism, and other threats to its security (JP 3-22). Foreign internal defense is an activity of irregular warfare and involves a comprehensive approach. The comprehensive approach includes all instruments of national power—diplomatic, information, military, and economic. Foreign internal defense is both a U.S. program and an operation. The program provides the authorities and permissions for the execution of operations. Together, the program and operations provide the United States with a capability that is neither enemy focused nor reactive in nature but oriented on proactive security cooperation. These activities shape the operational environment and prevent or deter conflict through military engagement with host nations, regional partners, and indigenous populations and their institutions.

2-24. Foreign internal defense is executed through unified action involving the synchronization, coordination, and integration of activities from governmental and nongovernmental entities within the operation to achieve unity of effort. The Department of State is normally the lead agency for execution of the programs with overall responsibility for the security assistance programs. The focus of assistance is to enable the host nation in anticipating, precluding, and, as a last resort, countering a threat. The lead military instrument in this collaborative environment may be a country team or a joint force commander. The Department of Defense provides the personnel and equipment to achieve program objectives. Foreign internal defense operations are executed best by avoiding deployment of large numbers of U.S. military personnel.

2-25. Army special operations Soldiers conducting or supporting foreign internal defense must be adaptive problem solvers and creative thinkers with the ability to work in a collaborative environment, building interagency and international partner capacity through a comprehensive approach. Foreign internal defense can be characterized as indirect support (indirect approach), direct support (direct approach), or combat operations; however, U.S. forces may simultaneously conduct some degree of all three forms of support (approaches) at different locations and times during operations. The approach may either involve economy-of-force and indirect approaches with Army special operations forces or the direct approach with the integration of special operations forces and conventional forces.

2-26. Foreign internal defense operations promote and protect U.S. national interests by influencing the threat and operational variables of political, military, economic, social, information, infrastructure, physical environment, and time through a combination of peacetime developmental, cooperative activities and coercive actions in response to crisis. Army forces, including special operations forces, accomplish stability goals through security cooperation. The military activities that support these operations are diverse, continuous, and often long-term. Their purpose is to promote and sustain regional and global stability. While foreign internal defense is an Army special operations forces core activity, stability tasks also employ Army special operations forces, in addition to Army forces, to assist civil authorities as they prepare for or respond to crises. Foreign internal defense is a broad program that covers a large range of activities. The primary intent is to help the legitimate host government address internal threats and their underlying causes. Commensurate with U.S. policy goals, the focus of all U.S. efforts is to support the host-nation internal defense and development program.

2-27. Foreign internal defense is not restricted to times of conflict. It is applied across the range of military operations that varies in purpose, scale, risk, and intensity of real operational environments. It can also take place in the form of training exercises and other activities that show U.S. resolve to and for the region. These exercises train the host nation to counter potential internal threats. The operations usually consist of indirect assistance, such as participation in combined exercises and training programs or limited direct assistance without U.S. participation in combat operations. These actions support the host nation in establishing internal defense and development programs. Foreign internal defense may require a security force assistance program to build host-nation capacity to anticipate, preclude, and counter threats or potential threats, particularly when the nation has not attained self-sufficiency and is faced with threats beyond its capability to handle. A supporting security force assistance program must emphasize internal defense and development. This emphasis helps the host nation address the root causes of instability in a preventive manner rather than reacting to threats.

> *Note:* ATP 3-05.2, *Foreign Internal Defense*, provides details on how Army special operations forces enable a host-nation's forces to maintain internal stability; to counter subversion, lawlessness, insurgency, terrorism, and other threats to security in their country; and to address the causes of instability averting failing state conditions.

MILITARY INFORMATION SUPPORT OPERATIONS

2-28. Military information support operations are a core activity which Psychological Operations forces are specifically organized, trained, and equipped to execute. Designated U.S. Army Special Operations Command and U.S. Army Psychological Operations units conduct specialized activities for commanders. Military information support operations are integrated in all Army special operations. Other core activities may support military information support operations as a line of operation to create specific psychological effects. Military information support operations can augment other capabilities, be the main effort, and are

Chapter 2

conducted across the range of military operations at the strategic, operational, and tactical levels of war. Military information support operations are conducted in conjunction with interagency activities to achieve U.S. national objectives. It is important not to confuse unintended psychological impact with planned psychological effects as part of military information support operations. While all military activities can have some degree of psychological impact, unless they are planned and executed specifically to influence the perceptions and the subsequent behavior of a target audience, they are not military information support operations. CJCSI 3110.05F, *Military Information Support Operations Supplement to the Joint Strategic Capabilities Plan,* and DODI O-3607.02, *Military Information Support Operations,* which provide requirements for permissions and authorities for military information support operations and restrict their execution to specifically trained forces, govern military information support operations. This restriction places responsibility on Psychological Operations Soldiers to advise commanders on psychological effects and makes the Army's Psychological Operations branch the Department of Defense's primary influence-focused force. In support of military information support operations or as separate activities, Psychological Operations units conduct activities to—

- Achieve psychological objectives in foreign audiences.
- Analyze and address psychological factors in the operational environment.
- Conduct influence activities as a core special operations capability.
- Influence activity across the range of military operations.
- Support other agency influence efforts (interagency/intergovernmental support).
- Support the countering of adversary information.
- Provide an important influence capability under mission command, synchronized through the fires warfighting function and integrated by the joint function, information.
- Conduct military and tactical deceptions.
- Train and advise host-nation forces on building organic influence capacity.
- Conduct precision influence targeting.
- Analyze targets and audiences within the operational environment.
- Analyze the communications environment within the operational environment.

2-29. In today's complex and rapidly evolving information environment, perceptions, decisions, and, ultimately, behavior are influenced by the psychological effects of actions and information. Emphasis on psychological objectives gives Psychological Operations Soldiers the responsibility to advise U.S. military commanders, ambassadors, or host-nation civilian and military leadership on the potential impact of messages and actions on targets and audiences. Military information support operations can be conducted unilaterally or in conjunction with economic, social, and political activities to limit or preclude the use of military force and to more effectively and efficiently use special operations forces for other high payoff missions. In some cases, the military objective may be relevant only in terms of the psychological effect. This emphasis on psychological effects has created a fundamental shift in the way Army special operations forces view military objectives from a planning standpoint. No longer can commanders look at just the physical aspects of an objective without taking into account the affected populations or targets and audiences in the area of operations. This shift affects all Army special operations forces operating in or through the information environment and increases the relevance of military information support operations as a core activity and capability to affect the environment.

PREPARATION OF THE ENVIRONMENT

2-30. *Preparation of the environment* is an umbrella term for operations and activities conducted by selectively trained special operations forces to develop an environment for potential future special operations (JP 3-05). Army special operations forces conduct these operations and activities in support of a combatant commander campaign plan to create conditions conducive to the success of military operations. Preparation of the environment operations and activities typically require a low-visibility posture and are planned using the principles of discreet, precise, and scalable.

2-31. Preparation of the environment includes operational preparation of the environment, advanced force operations, and intelligence operations. In the context of unified land operations and large-scale combat operations, preparation of the environment activities and operations conducted during the initial phase(s) of an operation or campaign help define the operational environment and prepare for the entry of forces and supporting governmental agencies.

2-32. *Operational preparation of the environment* is the conduct of activities in likely or potential areas of operations to prepare and shape the operational environment (JP 3-05). Combatant commanders leverage operational preparation of the environment to develop knowledge of the operational environment, to establish human and physical infrastructure, and to develop potential targets. Army special operations forces may use any combination of passive observation, area familiarization, site surveys, mapping the information environment, military source operations, developing nonconventional and unconventional assisted recovery capabilities, use of couriers, developing safe houses and assembly areas, positioning transportation assets, and cache emplacement/recovery to support operational preparation of the environment.

2-33. Advanced force operations are those operations that precede main forces into an area to locate and shape the operational environment for future operations against a specific adversary. *Advanced force operations* are operations conducted to refine the location of specific, identified targets and further develop the operational environment for near-term missions (JP 3-05). Advanced force operations encompass many activities including close-target reconnaissance; tagging, tracking, and locating; reception, staging, onward movement, and integration of forces; infrastructure development; and terminal guidance. Unless specifically withheld, advanced force operations also include direct action in situations when failure to act will mean loss of fleeting opportunity for success.

2-34. By operating in areas denied to larger conventional force elements, Army special operations forces enable decisive action by facilitating opportunities for conventional force maneuver commanders to seize, retain, and exploit the operational initiative. Advanced force operations may be decisive points in combat operations.

2-35. *Intelligence operations* are the variety of intelligence and counterintelligence tasks that are carried out by various intelligence organizations and activities within the intelligence process (JP 2-01). Intelligence operations include human intelligence activities (to include military source operations); counterintelligence activities; airborne, maritime, and ground-based signals intelligence; tagging, tracking, and locating; surveillance; and reconnaissance. Intelligence operations are critical for the conduct of operational preparation of the environment and advanced force operations.

SECURITY FORCE ASSISTANCE

2-36. *Security force assistance* is defined as the Department of Defense activities that support the development of the capacity and capability of foreign security forces and their supporting institutions (JP 3-20). Security force assistance encompasses Department of Defense efforts to support the professionalization and the sustainable development of the foreign security forces and supporting institutions of host countries, as well as international and regional security organizations. Security force assistance occurs across the range of military operations and within every phase of a military operation. Security force assistance activities are primarily executed to assist a host country to defend against both internal and transregional threats to security. Army special operations forces also conduct security force assistance to assist host countries defend effectively against external threats; contribute to coalition operations; or organize, train, equip, and advise another country's security forces or supporting institutions. The clear and expressed intent of any security force assistance activity is improving the capacity or capability of a foreign security force or its supporting institutions.

2-37. U.S. national security strategies and laws provide the foundation for foreign assistance. Policy and strategy determines where, when, why, and how the United States should invest in developing foreign security forces. Army special operations forces conduct security force assistance to support national objectives relating to developing the capacity and capability of foreign security forces.

2-38. Operationally, Army special operations forces plan security force assistance with a focus on understanding and then resolving the underlying problems facing the foreign security force vice merely resolving symptoms. At the tactical or execution level, Army special operations forces increase focus on achieving results. The execution of these activities may be indistinguishable from other tasks conducted during other special operations; however, the singular purpose of security force assistance plan, program, or effort is supporting the professionalization and the sustainable development of the capacity and capability of the foreign security force.

Chapter 2

SPECIAL RECONNAISSANCE

2-39. *Special reconnaissance* is defined as reconnaissance and surveillance actions conducted as a special operation in hostile, denied, or diplomatically and/or politically sensitive environments to collect or verify information of strategic or operational significance, employing military capabilities not normally found in conventional forces (JP 3-05). These actions provide an additive capability for commanders and supplement other conventional reconnaissance and surveillance actions. Special reconnaissance may include—

- Information on activities of an actual or potential enemy or secure data on the meteorological, hydrographic, or geographic characteristics of a particular area.
- Assessment of chemical, biological, radiological, and nuclear or environmental hazards in a denied area.
- Target acquisition, area assessment, and post-strike reconnaissance.

2-40. Special reconnaissance complements national and theater of operations intelligence collection assets and systems by obtaining specific, well-defined, and time-sensitive information of strategic or operational significance. It may complement other collection methods constrained by weather effects, terrain-masking, or hostile countermeasures. Selected Army special operations units conduct a human intelligence activity that places U.S. or U.S.-controlled "eyes on target," when authorized, in hostile, denied, or politically sensitive territory.

2-41. In an operational environment, the special operations and conventional command relationship may be that of supported and supporting, rather than tactical control or operational control. Using special operations forces with conventional forces by a joint force commander establishes interdependence and creates an additional and unique capability to achieve objectives that may not be otherwise attainable. Special reconnaissance enhances a joint force commander's situational awareness and facilitates staff planning of, and training for, unified action. However, such use does not mean that Army special operations forces will become dedicated reconnaissance assets for conventional forces. Instead, the joint force commander (through a subordinate special operations command element or theater special operations command) may task an element to provide reconnaissance information to conventional forces that may be operating within a joint special operations area. The joint force commander may also task an element on a case-by-case basis to conduct special reconnaissance within a conventional force area of operation. Special operations and conventional forces working within the same area of operations may develop formal or informal information-sharing relationships that enhance each other's operational capabilities.

2-42. Army special operations forces may employ advanced reconnaissance and surveillance sensors and collection methods that use indigenous assets. When received and passed to users, this intelligence is considered reliable and accurate and normally does not require secondary confirmation.

UNCONVENTIONAL WARFARE

2-43. *Unconventional warfare* is defined as activities conducted to enable a resistance movement or insurgency to coerce, disrupt, or overthrow a government or occupying power by operating through or with an underground, auxiliary, and guerrilla force in a denied area (JP 3-05.1). An **underground** is **a cellular covert element within unconventional warfare that is compartmentalized and conducts covert or clandestine activities in areas normally denied to the auxiliary and the guerrilla force (ADP 3-05).** An *auxiliary* is, **for the purpose of unconventional warfare, the support element of the irregular organization whose organization and operations are clandestine in nature and whose members do not openly indicate their sympathy or involvement with the irregular movement (ADP 3-05).** A *guerrilla force* is a group of irregular, predominantly indigenous personnel organized along military lines to conduct military and paramilitary operations in enemy-held, hostile, or denied territory (JP 3-05). Unconventional warfare operations, missions, and tasks are politically sensitive activities that involve a high degree of military risk and require distinct authorities and precise planning often characterized by innovative campaign design.

2-44. The term insurgency is used to describe both a group and the method and purpose of the group. *Insurgency* is defined as the organized use of subversion and violence to seize, nullify, or challenge political control of a region (JP 3-24). A *resistance movement* is an organized effort by some portion of the civil population of a country to resist the legally established government or an occupying power and to disrupt civil order and stability (JP 3-05). The slight differences in the terms can lead to misunderstanding and misconstruing the relationship between the two types of groups.

2-45. An insurgency group—
- Uses subversion and violence.
- Is made up of a portion of the country's population. It can also be an external organization.

2-46. A resistance movement—
- Does not have to use violence.
- Is made up of a portion of the country's civil population.

2-47. The United States conducts unconventional warfare in two primary ways in order to achieve strategic outcomes. The first is as a supporting line of operation to a larger military campaign or operation such as large-scale combat operations or a limited contingency operation. The second is as the strategic main effort to preempt or respond to aggression. Experiences in the 1980s in Afghanistan and Nicaragua proved that support for an insurgency could be an effective way of putting indirect pressure on the enemy. However, the costs versus the benefits of using these operations must be carefully considered before employment. Properly integrated and synchronized operations can extend the application of military power for strategic goals. Unconventional warfare can complement other operations by giving the United States opportunities to seize the initiative through preemptive covert or clandestine offensive action without an overt commitment of a large number of conventional forces.

2-48. Unconventional warfare has strategic utility that may alter the balance of power between sovereign states. Such high stakes carry significant political risk in both the international and domestic political arenas and necessarily require sensitive execution and oversight. The necessity to operate with a varying mix of clandestine and covert means, ways, and ends places a premium on intelligence of the operations area. In unconventional warfare, as in all conflict scenarios, U.S. military forces must closely coordinate their activities with interorganizational partners in order to enable and safeguard sensitive operations.

HOSTAGE RESCUE AND RECOVERY

2-49. Hostage rescue and recovery operations are sensitive crisis response missions in response to terrorist threats and incidents. Offensive operations in support of hostage rescue and recovery can include the recapture of U.S. facilities, installations, and sensitive material overseas (JP 3-05). Designated Army special operations forces may conduct hostage rescue and recovery operations independently or as part of larger conventional or unconventional operations or campaign. Hostage rescue and recovery operations may include close quarters battle tactics and techniques when the mission requires the recovery or capture of personnel or materiel.

2-50. Hostage rescue and recovery may also involve locating, recovering, and restoring to friendly control selected persons or materiel that are isolated and threatened in sensitive, denied, or contested areas. These missions usually result from situations that involve political sensitivity or military criticality of the personnel or materiel being recovered from remote or hostile environments. These situations may arise from a political change, combat action, chance happening, or mechanical mishap.

This page intentionally left blank.

Chapter 3

Command and Control Structures

A clear, responsive command structure designed to enable decentralized operations facilitates the execution of special operations across the conflict continuum. Unnecessary layering of a headquarters decreases responsiveness, decreases the amount of time available to plan, and creates opportunities to compromise operations. The command structure designed to enable the exercise of authority and direction over Army special operations units frequently dictates that echelons down to team may have to interact directly with joint forces, U.S. Embassy Ambassadors and staffs, interagency partners, and other unified action partners. Having a clear command structure facilitates shared understanding of operations with unified action partners and complements national security considerations.

The execution of command and control establishes the foundation for subordinate units to understand the commander's vision. These tasks enable the special operations force understanding of how the assigned mission nests with adjacent efforts, other operations, the commander's campaign plan, global operations, and national objectives. Through the product called commander's intent, commander's share their insight and direction to create a shared purpose and understanding and enable subordinates to achieve objectives without further orders in spite of unforeseen changes in the environment. The *commander's intent* is a clear and concise expression of the purpose of the operation and the desired military end state that supports mission command, provides focus to the staff, and helps subordinate and supporting commanders act to achieve the commander's desired results without further orders, even when the operation does not unfold as planned (JP 3-0). Army special operations unit commanders and the commander's they support place emphasis on mission command by providing a commander's intent that is clear, founded in shared understanding developed by analysis of the operational environment, and that expresses the purpose of the operation so their forces can execute decentralized operations.

UNITY OF EFFORT

3-1. Unity of effort and global integration requires coordination among government departments and agencies; nongovernmental organizations and intergovernmental organizations; and partners, allies, and nations in any alliance or coalition. Unity of effort, action, and command enables the synchronization, coordination, and/or integration of the activities of the governmental and nongovernmental entities with military operations to support a strategy. Global integration is the arrangement of cohesive military actions in time, space, and purpose, executed as a whole to address transregional, all-domain, and multi-functional challenges. Geographic, global, and functional combatant commanders are directly responsible to the President and the Secretary of Defense for the implementation of a strategy, U.S. policy, and the execution of assigned missions. The National Security Strategy, National Defense Strategy, and National Military Strategy—shaped to support the implementation of national security policies, and articulated through global, regional, and functional campaign plans—provide strategic direction for combatant commanders. In turn, combatant commanders design, plan, and execute campaign plans in accordance with this guidance as well as the Unified Command Plan, Contingency Planning Guidance, and Joint Strategic Campaign Plan. The Joint Strategic Campaign plan identifies which combatant commander is the coordinating authority for

Chapter 3

specific subordinate campaign plans. Specifying the coordinating authority facilities synchronization of U.S. operations between combatant commanders; with allies, coalitions, and multinational forces; and with nonmilitary organizations.

3-2. The Secretary of State is the President's principal foreign policy advisor. In the National Security Council interagency process, the Department of State is the lead agency for most U.S. Government activities abroad. For this reason, the Department of State plays a key role in special operations, to include validating requirements articulated by country teams; providing funding, authorities, and permissions through Title 22, United States Code; and providing a coordination point between Department of State activities, special operations, and other government agency activities within a region or country.

3-3. The United States maintains diplomatic relations with more than 180 foreign countries through embassies, consulates, and other diplomatic missions. The ambassador to a country is responsible to the President for directing, coordinating, and supervising official government activities and personnel in that country. These personnel include all U.S. military personnel not assigned to the geographic combatant commander or other designated U.S. military area commander. Protection and security of military personnel are a matter of significant interest. Often, specific agreements are required between the ambassador (also known as the Chief of Mission) and the geographic combatant commander. Special operations forces deployed to a particular country for any reason (exercise, operation, or security assistance) remain under the command and control of the geographic combatant command or a designated subordinate headquarters (normally the theater special operations command—see JP 3-22 for more information). Special operations forces do not operate in a combatant commander's area of responsibility or in an ambassador's country of assignment without providing prior notification and receiving permission to do so.

3-4. The Department of State requests for Army special operations capabilities usually originate from an ambassador, defense attaché, or security assistance organization chief, who passes the requests through the appropriate geographic combatant commander to the Chairman of the Joint Chiefs of Staff. The Chairman of the Joint Chiefs of Staff validates the requirement and ensures proper interagency coordination takes place. If the geographic combatant commander has available special operations forces and no restrictions exist on their employment, the geographic combatant commander can approve and support the request. If the required forces are unavailable, the geographic combatant commander requests forces through the Joint Chiefs of Staff to United States Special Operations Command.

3-5. Combatant commands requiring Army special operations forces to conduct planning, staff assistance visits, or other activities that do not require deployment orders or execution orders, request that assistance directly from United States Special Operations Command, which validates the requirement and sources it from its headquarters or a component(s). Executing units follow theater procedures for obtaining permission to enter the theater, obtaining a visa and country clearance for example, prior to departing home station.

3-6. Unity of effort is not attained without trust between all participants. Developing mutual trust and working partnerships between all of the various organizations and especially the key leaders and staff of host-nation and international actors are essential to the success of stability tasks—especially security cooperation. Trust begins at the personal level. Trusted Army professionals understand that it is the combined military expertise that their organizations bring—combined with their individual character, competence, and commitment in adherence to the moral principles of the Army Ethic—that make the greatest positive contribution to the Army special operations forces mission. Interpersonal relationships of key actors at the local, regional, and national levels, built on shared understanding, mutual respect, and personal trust, are most important and effective in accomplishing stability tasks and security cooperation efforts. Trust also contributes to the legitimacy of the operation and those critical personal relationships assist in achieving the desired end states.

THEATER OF OPERATIONS ORGANIZATIONS

3-7. The Army's special operations units, down to the team level, are often called upon to coordinate their activities with theater-level organizations. This section briefly describes those integrations from an Army special operations forces' perspective to provide an understanding of command and coordination relationships and responsibilities.

THEATER ARMY

3-8. Per Title 10, United States Code, and DODD 5100.01, the theater Army's primary role is that of the Army Service component command for the geographic combatant commander. The theater Army provides a regionally oriented, long-term Army presence for security cooperation, deterrence, and the capability to perform as a joint task force headquarters or a joint force land component command for a limited contingency operation. Army forces under the component commander of the combatant commander are attached to the Army Service Component Command, which exercises operational control of those forces until the component commander transfers operational control of those forces to a subordinate joint force commander, such as a joint task force commander. Army special operations forces are an exception to this as the theater special operations command typically exercises operational control of all special operations forces regardless of Service.

3-9. The theater Army staff includes positions for Psychological Operations personnel within the G-3 (assistant chief of staff, operations) staff section and positions for Civil Affairs personnel within the G-3 and G-9 (assistant chief of staff, Civil Affairs operations) staff sections.

3-10. To facilitate daily Title 10 responsibilities of the Army service component command, United States Army Special Operations Command may provide a special operations command and control element. This is a special operations element that is the focal point for the synchronization of special operations forces activities with conventional forces activities (JP 3-05). The special operations command and control element locates with the Army service component command and conducts liaison between the theater special operations command and the Army service component command/theater Army.

3-11. Regardless of the command and control arrangements of Army special operations units in the theater area of responsibility, as the senior Army sustainment headquarters for the theater Army, the theater sustainment command provides sustainment support to Army special operations forces. The 528th Sustainment Brigade (Special Operations) (Airborne) maintains staff coordination lines of communication with the theater special operations command, the theater sustainment command, and expeditionary sustainment commands to facilitate sustainment to Army special operations units. The 528th provides Army special operations forces liaison elements that co-locate with the theater sustainment command. Through these liaison elements, the brigade fulfills their responsibilities to identify Army special operations units' support requirements and to ensure that support requirements are established.

Note: Chapter 6 provides a more detailed discussion of Army special operations forces sustainment support.

THEATER SPECIAL OPERATIONS COMMAND

3-12. A theater special operations command is a joint component command subordinate to a combatant command. For example, Special Operations Command Pacific is a subordinate unified command of United States Indo-Pacific Command. United States Forces Korea is also a subordinate unified command of United States Indo-Pacific Command and has a subordinate functional special operations component, United States Special Operations Command Korea. As stated in previous paragraphs, geographic combatant commanders normally exercise operational control of special operations forces attached or assigned to them through the theater special operations command. Commanders of theater special operations commands are the principal special operations advisor to the geographic combatant commander. The commander, theater special operations command has the additional role of joint force special operations component commander when the component commander establishes a joint task force.

3-13. There are six theater special operations commands within the Department of Defense:
- United States Special Operations Command Africa.
- United States Special Operations Command Central.
- United States Special Operations Command Europe.
- United States Special Operations Command North.
- United States Special Operations Command Pacific.
- United States Special Operations Command South.

3-14. The theater operations command is the primary theater of operations organization capable of performing broad, continuous missions uniquely suited to special operations forces' capabilities. The commander has three principal roles:

- **Joint Force Commander.** As the commander of a subordinate unified command, the commander is a joint force commander. As such, the commander has the authority to plan and conduct joint operations as directed by the geographic combatant commander and to exercise operational control of assigned and attached commands and forces. The commander may establish task forces, such as a special operations task force, to plan and execute these missions.
- **Theater of Operations Special Operations Advisor.** The commander advises the geographic combatant commander and the other component commanders on the proper use of special operations. The commander may develop specific recommendations for the assignment of special operations units in the theater of operations and opportunities to integrate special operations into the combatant commander's campaign plan. A best practice to facilitate this advisory role is for the combatant commander to establish the commander, theater special operations command as a special staff officer on the combatant command staff. The commander, theater special operations command can then identify specific positions within his staff to support his role in this special staff effort.
- **Joint Force Special Operations Component Commander.** When directed by the combatant commander, the commander, theater special operations command assumes the role of a joint force special operations component commander within a joint force. It is incumbent on the geographic combatant commander to establish appropriate command relationships between special operations units, such as the theater special operations command, joint force special operations component, special operations joint task force, joint special operations task force, and special operations task force. The commander, theater special operational command's ability to function as the joint force special operations component commander while maintaining his role as the theater special operations commander is finite. When the span of control includes multiple joint special operations task forces, the geographic combatant commander establishes a special operations joint task force, commanded by a Major General (Army or Service equivalent) or Lieutenant General. The task force commander is capable of assuming the additional role of joint force special operations component commander.

COMMAND AND CONTROL

3-15. The organizations discussed in this section are used across the range of military operations. The authorities and responsibilities described in the preceding sections on Unity of Effort, Theater Army, and Theater Special Operations Command never change; they are prescribed in public law through United States Code. A commander's campaign plan relies on the capabilities provided by the entire Department of Defense as well as those capabilities provided by other government agencies to create unity of effort. Army special operations units execute a variety of activities, missions, and tasks following the principles of discreet, precise, and scalable. Army special operations forces execute these principles in over 160 countries (roughly 80% of the countries in the world) on a daily basis. U.S. military and civilian authorities should be familiar with the terms for a number of special operations elements in addition to those described as task forces. Table 3-1, page 3-5, depicts Army special operations command and control echelons used where task forces are required.

Command and Control Structures

Table 3-1. Army special operations command and control echelons

Echelon	Description	Command Posts	Command Level
Special Operations Joint Task Force (also called SOJTF)	A modular, tailorable, and scalable special operations task force. The 1st Special Forces Command is a scalable element organized to establish a Special Operations Joint Task Force. The task force provides integrated, fully capable, and enabled joint special operations forces to geographic combatant commanders and subordinate joint force commanders. Roles may include— • Supporting unified action through interorganizational cooperation for special operations conducted in support of a campaign or major operation. • Executing joint (or multinational, combined force) force special operations component responsibilities for the operation or theater. • Planning, integrating, and conducting all military operations in a designated area.	• Main command post (CP) • Operational CP • Support area CP • Mobile command group	Major General, Lieutenant General, or Service equivalent
Joint Special Operations Task Force (also called JSOTF)	The most versatile special operations task force. It is a scalable element organized around a single special operations group or regiment-sized unit. It is a joint task force composed of special operations units from more than one Service. Roles may include— • Serving as the joint (or multinational, combined force) task force special operations component in deterrence, crisis response, and limited contingency operations. • Serving as a joint task force headquarters for security cooperation, deterrence, crisis response, or limited contingency operation. • Serving as a subordinate, tactical headquarters in security cooperation, deterrence, crisis response, limited contingency operations, or large-scale combat operations.	• Main CP • Operational CP • Tactical CP • Support area CP • Early entry CP • Mobile command group	Colonel, Brigadier General, or Service equivalent
Special Operations Task Force (also called SOTF)	A scalable element organized around a single special operations battalion-sized unit. Special operations task forces operate under the operational control of a higher echelon special operations task force or a commander, theater special operations command. Roles may include— • Serving as a subordinate tactical headquarters across the range of military operations. • Serving as the primary, tactical-level special operations headquarters in security cooperation, deterrence, crisis response, and limited contingency operations.	• Main CP • Tactical CP • Early entry CP • Mobile command group	Lieutenant Colonel or Service equivalent
Advanced Operations Base (also called AOB)	A small, temporary base established near or within a joint operations area to command, control, and support special operations training or tactical operations. These bases are normally established by a scalable element organized around a single special operations company-sized unit. The base operates under the operational control of a higher echelon special operations task force. The base— • May be required to displace. • Extends the functionality of a special operations task force. • Provides a secure location to conduct training, support, and mission preparation.	• Tactical CP • Early entry CP	Major or Service equivalent

Special Operations Joint Task Force

3-16. A combatant commander, subordinate unified command commander, or a commander, joint task force establishes a special operations joint task force to support unified action through interorganizational cooperation for special operations conducted in support of a campaign or major operation. The established task force improves conventional and special operations forces integration, interoperability, and interdependence. It acts as a single headquarters to plan and coordinate all theater or joint operational area special operations. The special operations joint task force includes special operations capabilities from more than one Service and may include capabilities provided by special operations forces in other countries. The task force may have conventional force units assigned or attached to it to support or enable execution of specific missions. It provides command and control over multiple subordinate joint special operations task forces. To source the special operations joint task force, the combatant command requests forces through the global force management process.

Note: JP 3-05 provides information on special operations joint task forces from a joint special operations force perspective.

3-17. United States Army Special Operations Command is the sole United States Special Operations Command component designated as the force provider for the core of a deployable special operations joint task force headquarters. Headquarters, United States Army 1st Special Forces Command (Airborne), is the force that the Commander, United States Army Special Operations Command has designated to fulfill this role. As a special operations joint task force, the commander and staff must be able to exercise command and control over U.S. and partner nation special operations and conventional forces. The command must be able to integrate Service and partner nation representatives into its headquarters and may designate key positions of responsibility to be filled from an element that has significant forces assigned to the task force. As stated previously, the task force is capable of providing command and control over multiple joint special operations task forces (sourced by U.S. Naval Special Warfare Command or an Army special operations unit). In addition, the task force provides a capability to command and control military information support task forces, joint civil-military operations task forces, a joint special operations air component (typically sourced by U.S. Air Force Special Operations Command), and subordinate task forces consisting of U.S. and partner nations' conventional and special operations forces.

Joint Special Operations Task Force

3-18. A joint special operations task force consists of special operations forces from multiple Services. Several of these task forces may be conducting operations in a joint operational area or within the component commander's area of responsibility. The task force may provide command and control over several subordinate elements including a joint special operations air component and subordinate task forces. The commander can arrange his subordinate forces functionally, by mission, or by assigned area of operation. Operational art, operational design, and the planning process help commanders determine the best arrangement of their forces. Regardless of the special operations Service element that forms the core of a joint special operations task force headquarters or commands the task force, each Service providing forces retains administrative control of their forces. The senior officer in the theater exercises this control. When the representatives are outside the joint special operations task force, Service liaisons within the headquarters commandant element facilitate administrative control and support.

Note: JP 3-05 provides information on joint special operations task forces from a joint special operations force perspective.

3-19. The core of a joint special operations task force headquarters, sourced from the Army's special operations command, is provided by the Ranger Regiment commander and his staff or by a Special Forces Group commander and his staff. The Ranger Regiment is the most appropriate headquarters when the mission of the task force primarily requires direct action; short duration strikes and other small-scale actions conducted as a special operation in hostile, denied, or diplomatically sensitive environments and which employ specialized military capabilities to seize, destroy, capture, exploit, recover, or damage designated targets (JP 3-05). For all other missions the Special Forces Group is the appropriate headquarters.

Command and Control Structures

3-20. Special Forces groups have the capability to control diverse supporting tasks and provide a unique perspective of the operational environment, the joint special operations area, and the area of operations founded on their regional alignment. Special Forces groups and their subordinate units have decades of experience leading task organized Army special operations forces in missions ranging from supporting security cooperation and interagency support to participating in armed conflict. Joint special operations task force commanders may source key positions in their staff from the Services providing significant portions of their joint force.

SPECIAL OPERATIONS TASK FORCE

3-21. Special Forces battalions and Ranger battalions may form the core of the headquarters and operational elements of a special operations task force. A special operations task force may be subordinate to a joint special operations task force, under the operational control of the theater special operations command, or any command relationship structure that best enables their mission. There may be multiple special operations task forces in a theater operating under differing command and control structures conducting missions supporting military engagement through major operations. While these task forces are not joint headquarters, they may have a variety of capabilities attached to them to enable operations or to liaise with other units. Army special operations capabilities that may be included in the task force include Psychological Operations units, Civil Affairs units, special operations aviation units, and communications, intelligence, and sustainment units from the special operations sustainment brigade. These capabilities may also include U.S. or partner nation conventional or special operations forces and interagency elements ranging from individuals to teams. The commander augments his staff with appropriate special staff officers and liaison officers taken from attached and supporting assets to integrate and synchronize all activities.

3-22. In particular, Army conventional forces can provide critical capabilities to a special operations task force. The conventional force may be individual augmentees or may be an entire unit (typically company level to team). The capabilities they provide span the majority of military occupational specialties in the Army and are not limited to maneuver elements, such as infantry companies. Artillery, chemical, cyberspace, engineers, logistics, public affairs, medical, and strategic intelligence are all examples of capabilities that may be required.

ADVANCED OPERATIONS BASE

3-23. An *advanced operations base* is **a small, temporary base established near or within a joint operations area to command, control, and support special operations training or tactical operations (ADP 3-05).** The capabilities of the headquarters section/operational detachment–bravo of a Special Forces company provides the core element from which to configure an advanced operations base. Facilities are normally austere, and the base may be ashore or afloat. If ashore, the base may include an airfield or unimproved airstrip, a pier, or an anchorage. The advanced operations base is temporary but normally in a fixed location. Analysis of the operational environment and the mission will drive the need and readiness to displace the base. The Special Forces operational detachment–bravo exercises command of its organic Special Forces operational detachments–alpha as well as those additional Army special operations and conventional forces attached to the commander to support the mission.

3-24. An advanced operations base and the command element responsible for it receives support from a special operations task force located at another operations base. The advanced operations base extends the functionality of the higher echelon base by providing a secure location where the operational detachment–bravo can—
- Conduct mission specific training.
- Prepare forces for infiltration.
- Exercise command and control.
- Sustain a force during mission execution.
- Conduct debriefings.
- Provide operational, administrative, and sustainment support to uncommitted special operations forces.
- Develop, organize, train, equip, advise, assist, and direct indigenous or other partner forces.

Chapter 3

3-25. The organization and functionality of an advanced operations base requires an organic element such as an operational detachment-bravo and additional individual augmentees, enablers, and staff personnel that are identified during mission analysis. The advanced operations base organization and functions vary with the mission, as do the duration and scope of operations, security, communications, and support requirements.

SPECIAL OPERATIONS FORCES SUPPORT TO TASK FORCES

3-26. Joint task force commanders, their subordinate commanders, and their staffs should be familiar with the following special operations organizations within the task forces.

Joint Special Operations Air Component Commander

3-27. U.S. Special Operations Command's Air Force component is the Air Force Special Operations Command. Air Force Special Operations Command has a requirement to provide the commander and core staff of a joint special operations air component. The joint special operations air component is functional air component that plans and executes joint special operations air activities and coordinates conventional air support for special operations forces with the joint force air component commander or Service equivalent. The U.S. Army Special Operations Aviation Command's 160th Special Operations Aviation Regiment (Airborne) may operate under the operational control of a joint special operations air component commander. The regimental commander and the subordinate battalion commanders have the capability of being a joint special operations air component commander; however, their staffs are not manned and equipped to provide the core staff of a component command and require augmentation to do so.

3-28. A joint special operations air component is not always the most suitable command structure for Army special operations aviation units. The operational environment and mission analysis may find that efficiencies are created when these units are placed under the operational control of a joint special operations task force or special operations task force.

Special Operations Liaison Element

3-29. A *special operations liaison element* is a special operations liaison team provided by the joint force special operations component commander to coordinate, deconflict, and synchronize special operations air, surface, and subsurface operations with conventional air operations (JP 3-05). The special operations liaison element is located with the joint force air component or other air component as appropriate. This element includes enough personnel to place liaisons at each division of the joint air operations center, staff an internal operations cell, and maintain its special operations peculiar communications infrastructure. The director of the element serves as the special operations component's personal liaison to the joint force air component commander. In addition to the tasks described in the definition, the element ensures coordination of special operations in the joint force air component commander's air tasking order and airspace control order.

Special Operations Command and Control Element

3-30. As discussed in previous paragraphs, a special operations command and control element may be provided to the theater Army. Within a joint task force, the special operations component commander, or a subordinate special operations force commander, exchanges liaisons with those commanders that his forces operate in proximity with and to commanders whose operations the special operations force supports or affects. A special operations command and control element is a liaison solution the commander may elect to employ when the liaison responsibilities exceed the capacity of a few liaison personnel.

3-31. From the Army special operations force perspective, a special operations command and control element provides an essential capability to coordinate unilateral special operations with a conventional ground force headquarters and with a supported conventional force commander. A special operations command and control element is task organized and provided a tailored communications package. The personnel may include forces from Special Forces, Ranger, Psychological Operations, Civil Affairs, Sustainment, and Special Operations Aviation units. The element is normally collocated at corps level and above, with smaller liaison teams operating at division level and below; however, the scale of the operation may require an expansion of the capabilities provided at echelons below corps. The supported unit provides the special operations command and control element the required administrative and logistics support. The element is the focal point for synchronization with the conventional forces. At corps level, the element coordinates with the corps current operations integration cell, fires cell, and battlefield coordination detachment to deconflict targets and operations. It provides near-real time locations of Army special operations units and provides overlays and other data to the fires cell and the battlefield coordination detachment.

3-32. The land component commander or Army forces commander is normally assigned a large portion of the joint operations area land mass as his area of operations. With this assigned geographic space comes authority and responsibility to accomplish assigned missions within it. The land component commander becomes the supported commander for all operations within his area of operations. The joint special operations commander may be designated a supporting commander to this land component commander or Army forces commander for missions conducted in that specific area of operations. This supported and supporting relationship does not have to include command relationships—it can primarily focus on requirements for coordination of operations. Thus a higher echelon, such as special operations joint task force, can retain operational control of its subordinate elements while those elements coordinate their activities in a supported commander's area of operations so interdependence is maintained.

Note: JP 1 and JP 3-0 provide more information on the authorities and responsibilities of supported and supporting commanders.

3-33. As a supporting commander, the joint special operations task force commander would employ a special operations command and control element to facilitate his supporting commander's responsibilities to a ground force commander. The element remains under the operational control of the special operations commander. The special operations command and control element assists the joint special operations task force commander in fulfilling his supporting commander's responsibilities in several ways. It provides a positive means for the commander to ascertain the supported commander's needs. The element may provide a responsive reporting capability in those situations in which the special operations commander is tasked to answer the supported commander's information requirements. The element can exercise command of designated special operations units when the commander determines the need for such a command relationship to facilitate his supporting commander's responsibilities. The element can also provide a monitoring capability if the commander decides to transfer operational or tactical control of his executing unit to the supported commander—for example, the attachment of Special Forces operational detachments with operational or tactical control to an Army forces commander in order to improve that commander's ability to employ subordinate multinational forces. The special operations commander can attach these forces and pass control to the other commander with appropriate mission restrictions in accordance with his determination on the employment of those forces, such as "no reorganization of forces authorized" or "for use only in an advisory role with the designated multinational force."

Liaisons

3-34. A best practice is to exchange liaisons between higher, lower, supporting, supported, and adjacent organizations. To integrate fully with conventional and joint operations, Army special operations units must maintain effective liaison and coordination elements with all components of the force to synchronize effects created in the joint operational area. To support this effort, joint forces, conventional forces, and special operations units exchange a variety of liaison and coordination elements in addition to the special operations command and control element provided by the higher special operations echelon commander. They range in size from individual liaisons to small coordination elements. Whatever their size or location, these elements coordinate, synchronize, and deconflict missions in the other unit's area of operation.

3-35. Liaison and coordination elements ensure the timely exchange of necessary operational and support information to aid mission execution and to prevent fratricide, duplication of effort, disruption of ongoing operations, and loss of intelligence sources. They may help coordinate fire support, overflight, aerial refueling, targeting, deception, information, influence and other theater of operations issues based on current and future missions. These efforts are crucial in coordinating limited resources and assets and in maintaining unity of effort and the tempo of an operation or campaign. Special operations unit commanders may also establish or receive additional liaison and coordination elements with higher and adjacent units or other agencies, as appropriate.

Chapter 3

ARMY SERVICE COMPONENT COMMAND STRUCTURE

3-36. U.S. Special Operations Command is one of the nine combatant commands in the U.S. military's structure, with Military Department and defense agency-like responsibilities. Section 167, Title 10, United States Code and DODD 5100.01, *Functions of the Department of Defense and its Major Components*, task the command with performing the department-like functions of organizing, training, equipping, and providing combat-ready personnel for employment by geographic combatant commanders. The command designs, plans, and executes campaigns and global operations in accordance with guidance established in the Unified Command Plan, Contingency Planning Guidance, and Joint Strategic Campaign Plan. In addition, these documents direct the command to synchronize Department of Defense global campaign planning against terrorist networks.

3-37. U.S. Special Operations Command is a unified combatant command, but it also has authorities and responsibilities in common with the departments. The command's main responsibilities include programming and maintaining the Major Force Program-11 budget; developing special operations strategy, doctrine, and tactics; ensuring the interoperability of special operations forces; conducting planning; and commanding continental U.S.-based special operations forces. To accomplish these responsibilities, the command is comprised of five key subordinate organizations. Four are Service components, and one is a subordinate unified command. U.S. Army Special Operations Command is its Army Service component.

UNITED STATES ARMY SPECIAL OPERATIONS COMMAND

3-38. The Commanding General, U.S. Army Special Operations Command fills roles as the commander of U.S. Special Operations Command's Army component and the commander of an Army Service Component Command. Its mission, assigned by U.S. Special Operations Command, is to man, train, equip, educate, organize, sustain, and support forces to conduct special operations across the full range of military operations in support of joint force commanders and interagency partners to meet theater and national objectives. The commander is responsible for recruiting, organizing, supplying, equipping, training, servicing, mobilizing, demobilizing, maintaining, administering, supporting, educating, and preparing the readiness of assigned Special Forces, Ranger, Special Operations Aviation, Psychological Operations, and Civil Affairs units, including those forces temporarily under the operational control of other unified commanders. The commander exercises command of continental United States-based Regular Army special operations forces. He also oversees and evaluates continental United States-based Army National Guard special operations forces. The command is responsible for the development of unique Army special operations doctrine; tactics, techniques, and procedures; and materiel.

3-39. U.S. Army Special Operations Command consists of three component subordinate commands and one component subordinate unit manned with civilians and Regular Army and Reserve Component military personnel. The component subordinate commands (figure 3-1, page 3-11) are the 1st Special Forces Command (Airborne), the U.S. Army John F. Kennedy Special Warfare, and the U.S. Army Special Operations Aviation Command (Airborne). The component subordinate unit is the 75th Ranger Regiment (Airborne).

1ST SPECIAL FORCES COMMAND (AIRBORNE)

3-40. The 1st Special Forces Command (Airborne) organizes, equips, trains, validates, and deploys forces to conduct special operations across the range of military operations, in support of the Commander, U.S. Special Operations Command and other combatant commanders, U.S. Ambassadors, and other governmental agencies as directed. The commander and staff are trained and equipped to establish the core staff of a special operations joint task force in support of major operations (crisis response and limited contingency operations) and campaigns (large-scale combat operations).

3-41. The command (figure 3-2, page 3-11) consists of eleven component subordinate units manned with civilians and Regular Army and Reserve Component military personnel. The component subordinate units consists of five Regular Army (1st, 3d, 5th, 7th, and 10th) Special Forces Groups (Airborne) and two Army National Guard (19th and 20th) Special Forces Groups (Airborne), the 95th Civil Affairs Brigade (Special Operations) (Airborne), the 4th and 8th Psychological Operations Groups (Airborne), and the 528th Sustainment Brigade (Special Operations) (Airborne).

Command and Control Structures

Figure 3-1. United States Army Special Operations Command

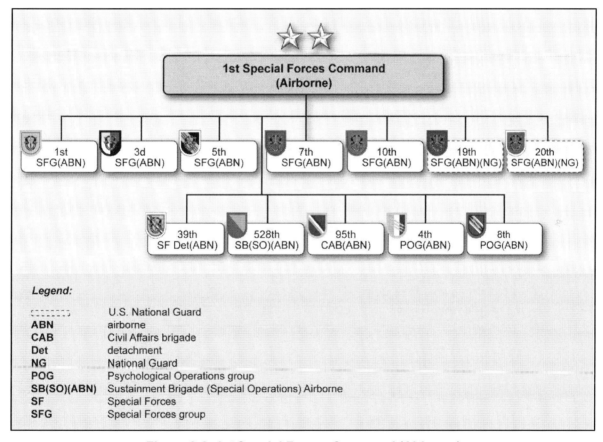

Figure 3-2. 1st Special Forces Command (Airborne)

Chapter 3

U.S. ARMY JOHN F. KENNEDY SPECIAL WARFARE CENTER AND SCHOOL

3-42. The U.S. Army John F. Kennedy Special Warfare Center and School is the proponent for the Civil Affairs, Psychological Operations, and Special Forces branches. It provides institutional training, doctrine, and personnel policy to support the aforementioned branches as well as for the Army special operations force as a whole. The Army has designated the U.S. Army John F. Kennedy Special Warfare Center and School as the Army's Special Operations Center of Excellence. A center of excellence is a designated command or organization within an area of expertise that executes assigned responsibilities.

3-43. Primary responsibilities as a center of excellence include planning, developing, executing, and assessing learning product development (to include leader development and education); remaining innovative and managing the resources provided; and managing personnel proponent requirements. Centers of excellence synchronize learning requirements across the operational, institutional, and self-development domains and use the analysis, design, development, implementation, and evaluation process to produce learning products. Centers of excellence maintain curriculum relevance, school accreditation, and development and sustainment of courseware.

3-44. The Special Operations Center of Excellence executes assigned responsibilities for U.S. Army Special Operations Command core functions; provides U.S. Army Special Operations Command the ability to develop and integrate doctrine, organization, training, material, leadership and education, personnel, facilities, and policy capabilities within and across the Army warfighting functions and the joint functions; and performs force modernization proponent responsibilities. Each functional component of the center ensures adherence to Army standards to enhance interoperability of conventional and special operations forces. The U.S. Army Special Operations Center of Excellence synchronizes and coordinates directly with the nine U.S. Army Training and Doctrine Command Centers of Excellence to achieve unity of effort, institutional agility, learning product visibility, and to improve product quality.

3-45. A validation process accredits the command to execute multiple functions, including the award of military occupational specialties, noncommissioned officer professional education, warrant officer professional education. In addition, this designation applies Army Enterprise Accreditation Standards in support of an Army learning institution.

3-46. The U.S. Army John F. Kennedy Special Warfare Center and School is the U.S. Army Special Operations Command's proponent for all matters pertaining to individual training, doctrine development, and all related individual and collective training material. It provides leader development, develops and maintains the proponent training programs and systems, and provides entry-level and advanced individual training and education for Civil Affairs, Psychological Operations, and Special Forces Soldiers. The U.S. Army John F. Kennedy Special Warfare Center and School (figure 3-3, page 3-13) consists of the following:

- Headquarters and Headquarters Company.
- Combined Arms Center Special Operations Forces Cell.
- Directorate of Training, Doctrine, and Proponency.
- Force Modernization Directorate.
- Special Forces Warrant Officer Institute.
- Noncommissioned Officer Academy.
- Special Warfare Medical Group.
- 1st Special Warfare Training Group.
- 2d Special Warfare Training Group.
- Office of the Civil Affairs Branch Commandant.
- Office of the Psychological Operations Branch Commandant.
- Office of the Special Forces Branch Commandant.

Command and Control Structures

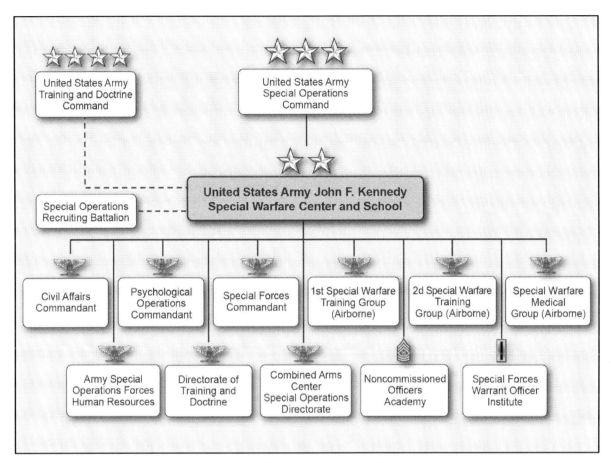

Figure 3-3. U.S. Army John F. Kennedy Special Warfare Center and School

ARMY SPECIAL OPERATIONS AVIATION COMMAND (AIRBORNE)

3-47. The command organizes, mans, trains, resources, and equips Army special operations aviation units to plan and execute special air operations in support of special operations. *Army special operations aviation* **are designated Active Component forces and units organized, trained, and equipped specifically to conduct air mobility, close air support, and other special air operations (ADP 3-05).** The command's special operations aviation units are specially organized, trained, and equipped to provide a Joint Force Special Operations Component Commander with the capability to infiltrate, resupply, and exfiltrate elements engaged in any of the special operations core activities. The command (figure 3-4, page 3-14) serves as the U.S. Army Special Operations Command aviation staff proponent and includes a technology applications program office, a flight detachment, a systems integration management office, a regimental organizational applications element, a training battalion, and the 160th Special Operations Aviation Regiment (Airborne).

75TH RANGER REGIMENT (AIRBORNE)

3-48. *Rangers* are rapidly deployable airborne light infantry organized and trained to conduct highly complex joint direct action operations in coordination with or in support of other special operations units of all Services (JP 3-05). The regiment can execute direct action operations in support of a combatant commander and can operate as conventional light infantry when properly augmented with other elements of combined arms. Its specially organized, trained, and equipped Soldiers provide a capability to deploy a credible military force quickly to any region of the world. It performs specific missions with other special operations units and often forms habitual relationships with these units. Its missions differ from conventional infantry forces' missions in the degree of risk and the requirement for precise, discriminate use of force. It uses specialized equipment, operational techniques, and several modes of infiltration and employment.

3-49. The headquarters of the 75th Ranger Regiment (figure 3-5) is similar to the headquarters of other Army brigade combat teams. In addition to exercising mission command for three Ranger infantry battalions and the Ranger Special Troops Battalion, the regiment headquarters may, when properly augmented, exercise operational control of conventional forces, logistics assets, and other special operations forces.

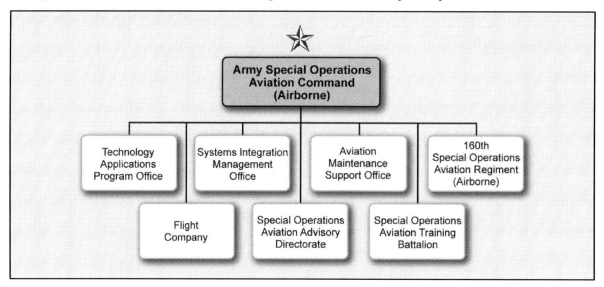

Figure 3-4. Army Special Operations Aviation Command (Airborne)

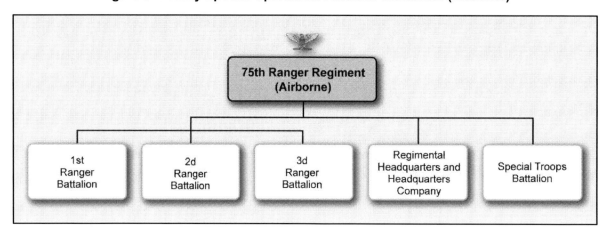

Figure 3-5. 75th Ranger Regiment (Airborne)

Chapter 4
Fires

Whether conducted unilaterally in support of a commander's campaign plan or as part of a joint, Army, or other Service effort, special operations can produce scalable lethal and nonlethal effects. Army special operations forces must be able to identify fires support required from the Army as well as the joint force. They must be able to integrate with the targeting processes that support the joint fires function and the Army fires warfighting function. In addition, they must be proficient in the various targeting methodologies in order to adapt the most suitable methodology for the specific special operation to be executed.

FIRES WARFIGHTING FUNCTION

4-1. The *fires warfighting function* is the related tasks and systems that create and converge effects in all domains against the adversary or enemy to enable operations across the range of military operations (ADP 3-0). Army fires systems deliver fires in support of offensive and defensive tasks to create specific lethal and nonlethal effects on a target. The fires warfighting function includes the following tasks:
- Deliver fires.
- Integrate all forms of Army, joint, and multinational fires.
- Conduct targeting.

4-2. Commanders use the analyses of the operational environment and their assigned area of operations in conjunction with the operations process to determine the effects required to achieve objectives. They use mission orders and the integrating process of targeting to obtain and/or provide the required capability in order to achieve that required effect. Army special operations commanders and planners make ethical, effective, and efficient decisions and take actions consistent with the moral principles of the Army Ethic. Employment of fires in decisive action requires the judicious use of lethal force that is balanced with restraint and tempered by professional judgment. Restraint requires the careful and disciplined balancing of the need for decisive combat action, security and protection, and the strategic end state (or overall combatant command mission). Army leaders should clearly understand how adherence to the Army Ethic provides a moral basis for decisive action and how it becomes a force multiplier in all operations.

Note: ADP 3-19, *Fires*, provides more information on the fires warfighting function.

TARGETING PROCESS INTEGRATION

4-3. Whether special operations are being conducted in support of large-scale combat operations or in support of the commander's campaign plan below the level of armed conflict, special operations forces consider political, military, economic, informational, and psychological effects on the enemy's capabilities, morale, and popular support base; on relevant populations; and on other relevant actors. Special operations units share the results of these considerations through the targeting process and through information and intelligence processes and systems in order to support shared understanding across the force. The knowledge produced by sharing considerations and analyses facilitates a comprehensive approach, identification of potential restricted targets, and the selection of the best capability to achieve the desired effects.

4-4. A comprehensive approach ensures that actions planned to achieve lethal and nonlethal effects are logical—from a chronological perspective and by considering that the effects created by each action may generate branches and sequels to the plan. Special operations forces' understanding of the human aspects of military operations incorporates a long-term perspective to any comprehensive approach, which is critical when supporting national objectives and enduring outcomes. The disciplined, ethical application of force frames the actions of trusted Army Professionals. Commanders and leaders, in formulating plans and orders, exercise restraint by considering nonlethal and lethal means to accomplish the mission.

4-5. The identification of potential, restricted targets and the selection of the best capability to achieve the desired effects both help mitigate unintended effects, to include unacceptable levels of collateral damage. Protection of noncombatant civilians and their property is a command responsibility and an essential element of the Army Ethic. Civilian casualty mitigation directly affects the success of the overall mission. In addition, the accountability, credibility, and legitimacy of a military operation, success of the overarching mission, and achievement of U.S. strategic objectives depend on the Army's ability to minimize harm to civilians. Failure to minimize civilian casualties can undermine national policy objectives and the strategic mission, while assisting the enemy. In addition, civilian casualties can incite increased opposition to friendly forces (ATP 3-07.6).

4-6. Nomination of restricted targets is an important part of any operation, but it can be substantially more important during operations that focus on achieving legitimacy and influence over a relevant population(s). Army special operations forces' analyses of the environment and how the environment affects the populations and groups within it result in thorough consideration of restricted target nominations, ranging from cultural artifacts, relevant actors, and infrastructure.

4-7. The selection of the best capability to achieve the desired effects first requires a detailed understanding of the environment. When determining the required effects to achieve an objective, the force considers all available capabilities. The use of lethal capabilities may be timely and feasible and its immediate effects are observable and collectible. However, planners must consider the potential enduring impact of unintended, primary, secondary, and tertiary effects as they may negate short-term desired effects. Nonlethal capabilities can often provide economy-of-force and create desired effects with less risk of collateral outcomes than lethal capabilities. However, the amount of time required to achieve the effect and the ability to collect measures of performance, measures of effectiveness, and indicators in order to assess the impact of the action, must be weighed against the need for timely achievement of the effect. The type of objective being supported and the risk associated with the use of a capability will factor into the selection of the most suitable capability to achieve the desired effect. The Army's special operations force provides both lethal and nonlethal capabilities, which provides commanders flexibility in achieving effects and objectives.

4-8. Special operations commanders and subordinate leaders are legally and morally responsible for their decisions and actions. They must work during their targeting planning process to take into consideration the civilian populace, noncombatants, friendly forces, and collateral damage when planning their operations. Special operations planners and leaders all have the legal and moral obligation to challenge a proposed mission if they believe it will violate the Law of War, Rules of Engagement, or the moral principles of the Army Ethic. Together, they must proactively plan and have the foresight to mitigate and reduce the risk of unintended effects, such as excessive collateral damage and negative psychological impacts on the civilian populace and other noncombatants—which create or reinforce instability in the area of operations. Improper planning could lead to severe consequences that adversely affect efforts to gain or maintain legitimacy and impede the attainment of both short-term and long-term goals.

SPECIAL OPERATIONS TASK FORCE FIRES CONSIDERATIONS

4-9. This section focuses on important considerations for joint task force commanders and their staffs and joint task force component commanders and their staffs concerning fires and special operations task forces. TC 3-05.5, *The Special Operations Task Force Planning and Operations Handbook*, provides detailed information on fires responsibilities and execution for special operations units conducting planning at the task force level.

4-10. When a special operations task force establishes a *joint fires element* (an optional staff element that provides recommendations to the operations directorate to accomplish fires planning and synchronization [JP 3-60]), it becomes the fires coordination link between commands. The joint fires element is responsible for planning joint fires and executing the targeting process within the special operations task force. The element is part of the current operations division. It consists of organic intelligence, sustainment, plans, communications, aviation, Special Forces, Civil Affairs, Ranger, and Psychological Operations personnel;

conventional force liaisons; representatives of attached units; and augmentees from the Services, multinational partner units, and government agencies that can achieve lethal and nonlethal effects, integrate into the targeting process, and advise on their organizations' capabilities. Examples of augmentees include United States Air Force air liaison officers and senior tactical air control party noncommissioned officers. These Air Force personnel advise commanders, through the operations directorate, on the capabilities and limitations of aerospace power. The air liaison officer works closely with the joint fires element in planning, requesting, and coordinating air support, to include joint close air support, air interdiction, intra-theater airlift, and combat search and rescue.

4-11. The exchange of liaisons further facilitates the coordination and synchronization of fires across a joint force. Liaisons are critical when a like position on a manning document was not created by the special operations task force, validated by higher or component headquarters, or filled by the force provider. Establishing liaisons between special operations and conventional force fires elements helps mitigate the tempo of armed conflict and the subsequent rapid information flow. Liaisons efficiently prioritize and pass critical information from one element to another vice relying on an element's capability to sift through vast quantities of data to retrieve the right information at the right time. Liaisons are a solution to mitigate interoperability challenges and to fulfill the need to inform elements of opportunities that may be fleeting.

4-12. When a higher headquarters or joint task force component requests special operations capabilities in order to create effects, the joint fires element conducts a *feasibility assessment*, a basic target analysis that provides an initial determination of the viability of a proposed target for special operations forces employment (JP 3-05). This assessment can require deliberate planning and analysis—for example when special operations capabilities are called upon to support or defeat a resistance movement—or it can be rapid, for example during the execution of a targeting board. In either case, the feasibility assessment includes answering the following questions:
- Is this an appropriate mission or task for special operations forces?
- Does the mission/task support the joint force commander's campaign or operation?
- Is the mission or task operationally feasible?
- Are the required support and execution resources available?
- Do the expected outcomes justify the risk to the force/command/geographic combatant commander, nation?

4-13. There are special operations elements that are external to the special operations task force. A special operations command and control element is established to coordinate unilateral special operations with a conventional force headquarters (most often a ground force) or, if the special operations task force is a supporting command, to facilitate the supporting commander's responsibilities. Special operations command and control elements are usually established at corps echelons and higher. These elements may be a permanent organization attached to a regionally aligned headquarters such as a theater Army or a temporary organization established for the duration of operations. The special operations command and control element tracks all operations in the environment and incorporates the coordination and synchronization made between fires cells/elements into the larger operations picture. The joint force special operations component commander establishes a special operations liaison element to coordinate, deconflict, and synchronize special operations with conventional air operations. The special operations liaison element is typically located at the joint air operations center or appropriate Service air component command and control center. The liaison element coordinates target nominations to the joint integrated prioritized target list and special operations in the air tasking order and the airspace control order.

4-14. A *time-sensitive target* is a joint force commander-validated target or set of targets requiring immediate response because it is a highly lucrative, fleeting target of opportunity or it poses (or will soon pose) a danger to friendly forces (JP 3-60). The joint force commander establishes guidance on procedures for coordination, deconfliction, and synchronization among components in a theater of operations or joint operations area. The special operations task force supports time-sensitive target prosecution, to include reconnaissance, surveillance, terminal guidance and control of weapons systems, direct action, and exploitation and amplification of the action for psychological effect. The joint fires element within the special operations task force integrates support to time-sensitive target prosecution through the special operations liaison element at the joint air operations center and/or the special operations command and control element at the corps or Marine expeditionary force.

Chapter 4

4-15. The integration of air-ground systems is a critical requirement to support joint fires. A special operations task force must be prepared to integrate the special operations air-ground system with the theater air-control system, Army air-ground systems, and other Service air-control systems separately or in combination. Likewise, these organizations must be capable of receiving and facilitating integration, and the theater Army/Army service support command may have responsibilities to facilitate integration as part of setting the theater. When the theater air control system, Army air-ground system, composite warfare commander/Navy tactical air control system, Marine air command and control system, and the special operations air-ground systems are all integrated, the combined system is referred to as the theater air-ground system. The theater air-ground system encompasses organization, personnel, equipment, and procedures.

> *Note:* JP 3-03, *Joint Interdiction*; JP 3-09.3, *Close Air Support*; and JP 3-30, *Command and Control of Joint Air Operations*, provide more information on the theater air-ground system and each Services' component system.

4-16. A *joint terminal attack controller* is a qualified (certified) Service member who, from a forward position, directs the action of combat aircraft engaged in close air support and other offensive air operations (JP 3-09.3). The joint terminal attack controllers are the forward element of the theater air-ground system, and they must be organized, trained, and equipped to operate within that infrastructure. Air Force and special operations unit joint terminal attack controllers work together to provide special operations task force units with the ability to control aircraft in support of operations.

4-17. An air support operations center is subordinate to the joint force air component commander's joint air operations center. The joint air component commander or Air Force Service component command may provide a joint air coordination element to a special operations task force in order to create an air support operations center capability at the task force and to provide expedient access to the Joint Air Request Network. The joint air coordination element structure resembles an air support operations squadron staff and includes multiple fighter duty officers and senior fighter duty and tactical air control party noncommissioned officers. The establishment of this capability at the special operations task force facilitates decentralized and remote operations by providing responsive support to special operations units operating in deep or denied areas.

Chapter 5

Intelligence

The intelligence enterprise consists of interconnected intelligence networks and nodes from the national to the tactical level. The enterprise is flexible and responsive enough to support special operations unit commanders and their forces as they plan for and execute a broad range of activities. This flexibility and responsiveness allows special operations units to generate intelligence, which senior military and civilian leaders use to inform decisions.

OVERVIEW

5-1. Special operations commanders depend on the intelligence process to provide accurate, detailed, and timely support to their forces and supported commanders. Special operations commanders drive the intelligence process by articulating their priorities for the intelligence effort. At every echelon, the unit's senior intelligence officer (intelligence directorate of a joint staff [J-2]/assistant chief of staff, intelligence [G-2]/intelligence staff officer [S-2]) is responsible for satisfying the commander's requirements by planning, directing, and coordinating for the production of intelligence and the execution of intelligence operations. Commanders and the Army special operations forces they lead rely upon their organic intelligence staff sections and intelligence units to fulfill intelligence requirements. In order to fulfill these requirements, these intelligence elements must be able to integrate with the theater special operations command J-2, the theater of operations J-2 and joint intelligence center, the theater intelligence brigades, the U.S. Special Operations Command joint intelligence center, and the 389th Military Intelligence Battalion (Special Operations) (Airborne). Integrating with these intelligence organizations ensures that requirements exceeding their organic capability or capacity are fulfilled and that special operations information and intelligence products are available to the joint force.

SPECIAL OPERATIONS INTELLIGENCE CRITERIA

5-2. Special operations missions are both intelligence-driven and intelligence-dependent. Intelligence products developed for these units must be detailed, accurate, relevant, predictive, and timely. These intelligence products enable commanders to identify and assess potential courses of action; plan operations; properly direct their forces; employ ethical, effective, and efficient tactics and techniques; and implement protection. For example, infiltrating into a hostile environment to conduct a noncombatant evacuation operation in a dense urban environment requires precise information about structures, elements of infrastructures, the surrounding populations and the location(s) of persons to be evacuated. National- and theater-level intelligence products are often required at a lower echelon than is normally associated with support to conventional operations. Special operations missions may also require near-real-time dissemination of intelligence directly to the lowest echelon—the force conducting actions on the objective.

5-3. Special operations intelligence support requirements are mission- and situation-dependent, largely driven by diverse and unique operational environments. These forces are tasked to address country- and region-specific challenges, as well as transregional, all-domain, and multifunctional challenges. Therefore, intelligence support requires a thorough understanding of special operations requirements at the tactical level and the integration of intelligence products from across the operational environments and geographic combatant commands.

Chapter 5

5-4. The following variables can affect intelligence support:
- Environment category (hostile, permissive, unknown).
- Multinational, combined, joint, or unilateral operations.
- Force composition.
- Maritime or land-based operations.
- Mission duration.
- Availability and interoperability of command, control, and intelligence system elements and architecture.
- Adversary capabilities, objectives, and operational approach.
- Connectivity to agencies outside the operational environment or area of operations.

NATIONAL-LEVEL INTELLIGENCE SUPPORT

5-5. The diverse capabilities of the intelligence community are important sources of intelligence for Army special operations units. Intelligence community sensors have a depth and breadth of coverage that allows them to see into denied or hostile areas where special operations units operate. Consequently, they provide unique and critical information and intelligence that may not be accessible from organic special operations intelligence formations. Army special operations units can often rely upon such intelligence community collectors and analytic teams to cover areas of interest early in crisis or contingency situations, when political sensitivities are high and when these forces are the first, or only, military forces committed. Often, the agencies that make up the intelligence community come together to form an interagency intelligence center that supports, works parallel to, or is integrated with the geographic combatant commander's joint intelligence center. Examples of collaborative capabilities include the Federal Bureau of Investigation's expeditionary forensic lab to liaison officers in the joint intelligence center that facilitate shared intelligence efforts.

5-6. An important element of the intelligence community is the Office of the Director of National Intelligence, which promotes the integration of a coalition of sixteen government agencies and organizations, which include the Defense Intelligence Agency, the National Security Agency, the National Geospatial-Intelligence Agency, the Federal Bureau of Investigation, and the Central Intelligence Agency. These named agencies are collectively referred to as the 'Big 5' within the intelligence community and are the primary intelligence community members that support special operations forces. Army special operations units can have embedded National Geospatial-Intelligence Agency analysts to provide additional support. These organizations, along with the Service intelligence organizations, support task force operations in each theater. They provide substantive intelligence assets, dedicated communications connectivity, personnel augmentation, and counterintelligence support. The focal point for national-level intelligence support to theater operations is through the supported theater joint intelligence center or joint analysis center from the National Military Joint Intelligence Center.

THEATER INTELLIGENCE

5-7. The theater special operations command J-2 is the intelligence focal point for special operations. The J-2—
- Ensures intelligence products are available to support each special operations command mission tasking.
- Coordinates with and relies on their respective theater joint intelligence center or joint analysis center (United States European Command) and Service organizations to collect, produce, and disseminate intelligence to meet needs.
- Validates, reconciles, consolidates, and prioritizes requirements to optimize collection and production efforts.
- Coordinates joint special operations intelligence operations and the production and dissemination of target intelligence packages to support operations targeting.
- Directs subordinate units to collect and report information supporting the theater special operations command's intelligence requirements.

5-8. The J-2 coordinates with the theater special operations command communications system directorate of a joint staff (J-6) to obtain secure (sensitive compartmented information) voice and data communications with subordinate, supporting, and supported units. In some missions, a special operations joint task force or joint special operations task force is established; and in those cases, the task force J-2 functions in the same manner as the theater special operations command J-2.

THEATER OF OPERATIONS JOINT INTELLIGENCE CENTER

5-9. The theater joint intelligence centers (or joint analysis center under United States European Command) are the primary area of responsibility all-source analysis and production organizations. National intelligence agency representatives are integrated into the joint intelligence center—augmenting its analytical and production capability. The center provides much of the intelligence production agency support needed for target intelligence packages required for missions identified through the targeting process. The center should fully integrate military information support operations and Civil Affairs operations information into its all-source analytical and production effort.

5-10. The center normally locates near the joint operations center. The center serves as the focal point for operational and intelligence support to crisis or contingency operations. It is also the primary theater of operations interface with the National Military Joint Intelligence Center.

UNITED STATES SPECIAL OPERATIONS COMMAND JOINT INTELLIGENCE CENTER

5-11. The U.S. Special Operations Command joint intelligence center provides complementary special operations-peculiar intelligence support to all theater special operations commands upon request of the theater of operations joint intelligence center. Specifically, its mission is to provide timely analysis, production, and dissemination of all-source intelligence relating to special operations and special operations core activities, to include—
- Combatant commanders.
- Theater special operations commands and supported commands.
- U.S. Special Operations Command's component commands and subordinate units.

5-12. U.S. Special Operations Command provides intelligence support to theater special operations commands from its joint intelligence center and from deployable intelligence support packages. The command deploys tailored packages of personnel, systems, and equipment to combatant commanders for direct support to theater of operations special operations forces. The personnel, systems, and equipment in these tailored packages achieve access to general military intelligence databases focused on the area of operations, plus other operational needs.

FUSION CENTER

5-13. Fusion centers improve dynamic operational support by integrating the mission command approach together with focused analysis and other partners within a single centralized entity. A fusion center is an ad hoc collaborative effort between several units, organizations, or agencies that provide resources, expertise, information, or intelligence to a center with the goal of supporting the operations of each contributing member. They are primarily designed to focus collection and promote information sharing across multiple participants within a set geographic area. These centers are not operations centers. Commanders at various echelons create fusion centers to manage the flow of information and intelligence; focus information collection to satisfy information requirements; and to process, exploit, analyze, and disseminate the resulting collection.

Note: FM 2-0, *Intelligence*, Appendix C, provides more information.

SPECIAL OPERATIONS INTELLIGENCE ARCHITECTURE

5-14. The flow of intelligence information and coordination, requests, tasking, production, dissemination, resourcing, and support require flexible and responsive structures. These structures are deliberately planned and then codified in mission orders and other authoritative documents.

Chapter 5

5-15. Figure 5-1, page 5-5, structurally depicts the intelligence flow in the context of Army special operations forces conducting daily operations in support of a combatant commander campaign plan. Figure 5-2, page 5-6, depicts the intelligence flow in the context of large-scale combat operations. In either case, a two-star level special operations command's J-2 provides the through point between subordinate intelligence elements and higher echelon intelligence elements. The 389th Military Intelligence Battalion (Special Operations) (Airborne) may provide additional support to this command, its deployed subordinate elements or both. The nature of the battalion's support relationship is established in military orders.

5-16. U.S. Special Operations Command and the geographic combatant commands it supports have agreements that authorize direct liaison between U.S. Army Special Operations Command operational elements and the supported theater special operations command for operational and exercise purposes. Organic S-2 organizations provide intelligence support to Civil Affairs, Psychological Operations, Ranger, Special Forces, special operations aviation elements, and special operations sustainment elements.

5-17. Special Forces Groups and the 75th Ranger Regiment have organic collection capabilities organized into Military Intelligence companies and detachments. These multidiscipline collection capabilities enable organic collection, when task organized and deployed, in support of special operations. In support of this mission, special operations intelligence staffs process requests for information, tailor, and disseminate products produced at the joint force, combatant command, and national levels for special operations. They combine this tailored intelligence with organic intelligence collection and tactical information collected by subordinates to develop all-source intelligence and products for the commander, staff, and operators. Critical to this process is the translation of operational requirements, articulated by tactical subordinates, into the information collection plan. Joint and Service component interoperability of intelligence systems is critical to successful information sharing and collaboration across the intelligence enterprise. Army special operations forces intelligence elements typically have a mix of Army and special operations-specific intelligence systems to enable this collaboration. Detailed theater intelligence architecture planning and coordination, for each deployment or operation, is essential for success.

5-18. The 389th Military Intelligence Battalion (Special Operations) (Airborne) is the nexus for continental United States-based intelligence support, integrating the efforts of each U.S. Army Special Operations Command component. The military intelligence battalion, with the component intelligence staffs, coordinate with their staff counterparts and their units' communications elements to ensure continuous intelligence reach through the Joint Worldwide Intelligence Communications System. By ensuring interoperability within theater of operations systems and access to national and Service-unique information sources, the commander should have the required intelligence support for the full range of missions.

INTELLIGENCE PROCESSING, EXPLOITATION, AND DISSEMINATION

5-19. Army special operations forces are a key enabler in decisive, shaping, and sustaining operations by conducting activities that result in obtaining actionable intelligence. The processing, exploitation, and dissemination of such intelligence assists commanders in determining the appropriate capability in preparing the force to conduct activities against threat networks and other related hybrid threats. The activities could be unilateral, in conjunction with indigenous forces, or with joint conventional forces.

5-20. Army special operations capabilities are managed to be balanced and sustainable across operations, activities, and tasks to accomplish strategic end states in support of theater operations. The balanced approach includes the ability to combine team special operations activities with organic enabling capabilities of intelligence operations, surveillance, reconnaissance, and rotary-wing aircraft in support of unified land operations. The synchronization and integration of processing, exploitation, and dissemination with sensors and assets focuses on priority intelligence requirements, while answering the commander's critical information requirements. Processing, exploitation, and dissemination activities facilitate timely, relevant, usable, and tailored intelligence.

Intelligence

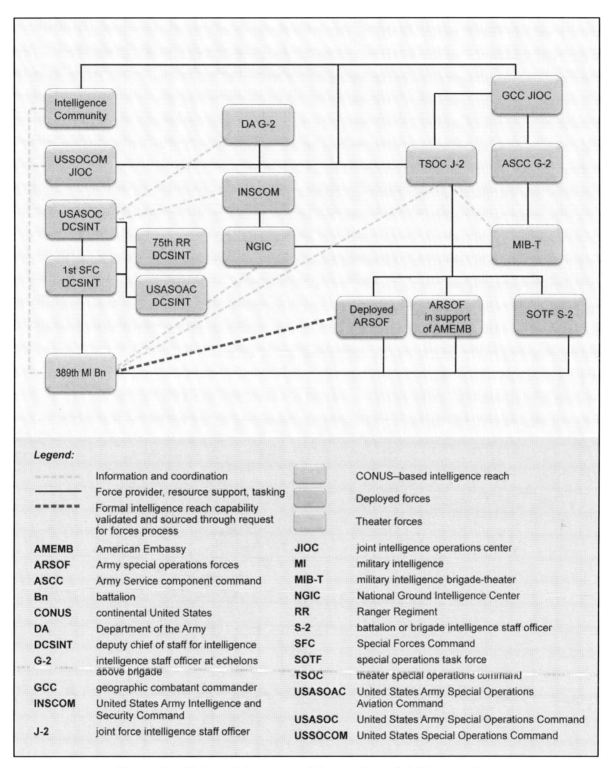

Figure 5-1. Notional Army special operations intelligence flow
for combatant commander daily operations

Chapter 5

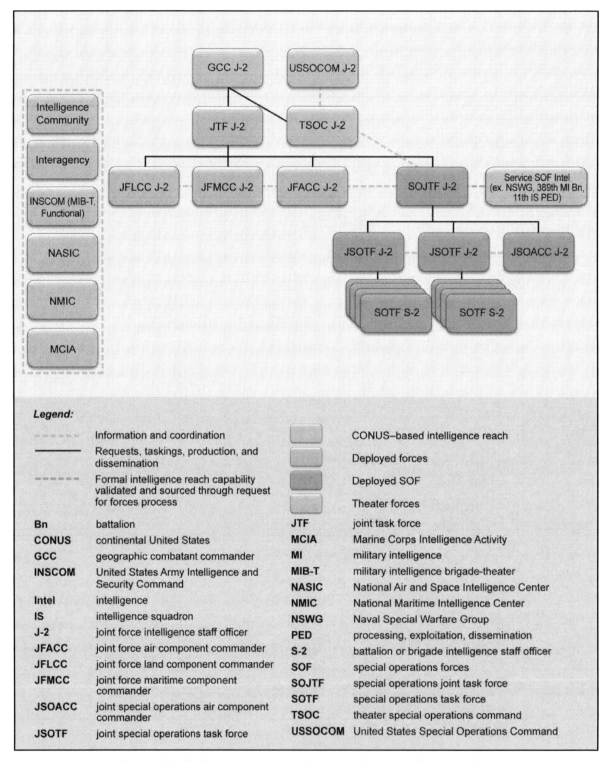

Figure 5-2. Notional Army special operations intelligence flow for large-scale combat operations

5-21. In joint doctrine, processing, exploitation, and dissemination is a general concept that facilitates the allocation of assets to support intelligence operations. Under the processing, exploitation, and dissemination concept, planners examine all collection assets and then determine if allocation of additional personnel and systems are required to support the exploitation of the collected information. These enablers are distinct from intelligence collection systems and all-source analysis capabilities. Intelligence processing, exploitation, and dissemination activities are prioritized and focused on intelligence processing, analysis, and assessment to quickly support specific intelligence collection requirements and facilitate improved intelligence operations. Because of the complexity of this task, the G-2 or S-2 plans the intelligence processing, exploitation, and dissemination portion of the intelligence architecture and then advises the commander on prioritizing and resourcing the supporting activities. A thorough assessment of intelligence processing, exploitation, and dissemination activities requires an understanding of the capabilities and the requirements for many different supporting systems and personnel from across the intelligence enterprise.

Note: ADP 2-0 includes more information on intelligence processing, exploitation, and dissemination.

This page intentionally left blank.

Chapter 6

Sustainment

The Army special operations sustainment infrastructure is deliberately limited when compared to a conventional force infrastructure of similar echelon. In order to conduct discreet, precise, and scalable special operations, Army special operations units rely on external sustainment capabilities provided or facilitated by the Theater Army, joint force commander, supported unit commander, partner nation, U.S. Embassy, or an indigenous source(s). Unit sustainment organizations coordinate with these external sources for logistics, personnel services, and health service support that exceed their organic capabilities. The planning and execution of sustainment for special operations must be nested within the joint force commander's concepts of operation and sustainment and tailored to interface with the theater of operations sustainment structures. To be effective, Army and special operations planners must understand the sustainment organizations' operational concepts, the basic principles of sustainment, and the sustainment warfighting function.

SUSTAINMENT BRIGADE

6-1. U.S. Army Special Operations Command is assigned one sustainment brigade, the 528th Sustainment Brigade (Special Operations) (Airborne), that is deployable in support of Army special operations force led task forces. The 528th provides the capability to set the operational-level sustainment conditions to enable special operations. It accomplishes this by coordinating requirements with the theater Army and theater special operations command and ensuring the theater sustainment structure is responsive to those requirements. The nature of special operations and the environments in which they are conducted requires the 528th commanders, staffs, and Soldiers to be agile, adaptive, and innovative stewards of the Army profession. They should creatively and competently support special operations and accomplish their missions in the right way: ethically, effectively, and efficiently. This way, the 528th allows continuous sustainment operations for any given special operations to persevere and succeed.

6-2. The 528th is unique in comparison to other Army sustainment brigades in that it focuses at the operational level for sustainment planning and synchronization, and it is trained, manned, and equipped to deploy small, modular teams to conduct coordination and participate in theater-level planning. The brigade can serve as the senior sustainment unit in a joint operations area. With augmentation, it can establish theater-opening and intermediate staging base operations with tailored multifunctional Army sustainment enablers. The brigade provides Role 2 medical support including expeditionary initial resuscitation and stabilization, limited holding, critical care patient staging, and en route critical care to deployed Army special operations forces. The special operations resuscitation team can provide reinforcing support and integrate with forward surgical teams.

PLANNING

6-3. Planning sustainment support is vital. The type of environment and operation, phase of the campaign or operation, the deployment sequence, unit basing strategy, and the assigned operational area are all factors that must be considered during the planning and operations processes.

6-4. A robust sustainment system that builds up over time into a mature logistics infrastructure characterizes a joint operational area in protracted armed conflict. The command designated to provide sustainment to the joint operational area must fulfill the validated requirements identified to sustain special operations. The theater Army is responsible for providing sustainment. The Army Service Component Command's aligned

Chapter 6

theater sustainment command is overall responsible for providing sustainment, but it may execute this responsibility through an expeditionary sustainment command or a sustainment brigade. Special operations sustainment planners must then concentrate on—

- **Initial Entry.** They must determine the type of sustainment required, the number of days of accompanying supplies based on the time-phased force and deployment list, and the unit's basing needs.
- **Buildup and Integration.** They must coordinate and integrate special operations sustainment requirements with the theater of operations support system before time-phased force and deployment list closure and as it continues to mature. In some cases, the theater of operations sustainment infrastructure never achieves full maturity.
- **Redeployment.** As units begin redeployment, the Army service component command ensures the remaining sustainment units and infrastructure are tailored to meet the requirements of units still executing operations in the joint operational area.

6-5. Each operation is unique and requires mission-specific analysis that develops a tailored sustainment force. Joint, interagency, intergovernmental, and multinational activities add complexity to the sustainment system. Special operations may be conducted in locations that make theater organic sustainment infrastructure support unfeasible. In these cases, the theater sustainment organizations facilitate, but do not provide, required sustainment support. In order to acquire support in these situations, planning efforts must incorporate a mix of theater assigned personnel; for example, geographic combatant command sustainment planners, theater sustainment command planners, and sustainment planners from the theater special operations command.

6-6. All sustainment operations constantly strive to maintain units at a desired level of readiness. To maintain the desired level, planners must—

- Coordinate with the theater sustainment command, expeditionary sustainment command, or Army sustainment brigades, as appropriate.
- Maximize the use of existing fixed facilities.
- Limit sustainment requirements to mission essentials within acceptable risk.
- Minimize the handling of supplies.
- Concentrate maintenance on returning major end items to service.
- Rely on air lines of communications for rapid resupply.
- Anticipate high attrition of supplies while performing missions in denied areas.
- Identify to the supporting sustainment headquarters, as early as possible, those items that require special logistics arrangements.
- Make maximum use of host-nation support, including indigenous and other partner resources.
- Conduct threat assessment.
- Conduct risk assessment.

6-7. During planning and preparation, the theater special operations command and the executing special operations unit can identify the requirements in operation plans and concept plans, including requirements down to the operator level, based on established planning assumptions. In this way, the Army special operations forces liaison elements can provide operational level special operations sustainment planning support. Including these units into planning efforts facilitates the Army special operations forces liaison element's ability to coordinate for Army common-user logistics and services through the Army service component command and for special operations forces-peculiar item supply, sustainment, and maintenance through the United States Special Operations Command and/or the United States Army Special Operations Command. Changes in the operational environment, the validity of assumptions, force structure, and capabilities may require modifications to planned requirements and their supporting sustainment plan to capture new requirements and subsequent changes to the sustainment plan.

6-8. The theater special operations command may request Army special operations forces to assist the planning process by participating directly in the theater-level planning process and conducting assessments or site surveys. The latter two activities can also serve theater Army preparations.

Sustainment

6-9. The use of assessment teams to support planning is not always practical. During the planning process, the theater special operations command and the Army service component command staffs must anticipate the combatant command's support requirements to facilitate the theater Army's efforts to set the theater. When required, special operations sustainment units can deploy advance party personnel to assist with reception of special operations forces and to establish access for those forces to the theater support structure.

6-10. The geographic combatant commander establishes command relationships for Army special operations forces operating in theater. The theater Army has Title 10, United States Code, responsibility—regardless of command relations within the combatant command—to provide administration and support to deployed Army forces. When directed by the geographic combatant commander, the theater Army supports and sustains other services and multinational special operations forces.

6-11. The theater Army conducts reception, staging, onward movement, and integration and follow-on support and sustainment of theater of operations Army forces including those in intermediate staging bases. The 528th may be required to conduct reception, staging, onward movement and integration activities for special operations forces during initial entry into an area of operations, if theater army forces are not available.

6-12. The principles of special operations (discreet, precise, and scalable) and the environments in which those operations are conducted generate consistent conditions that should receive deliberate consideration by sustainment planners. Special operations units operate in isolated and austere locations with constraints on the visibility or public affairs posture of those operations and the forces conducting them. This combination of conditions makes logistics distribution challenging and dictates that the most suitable capability is tasked to support Army special operations requirements. Sustainment planners must consider each Services' capability, contracting options, and special agreements, such as cross service, inter-organizational, or cross government. In addition, the maintenance of both Army equipment and special operations peculiar equipment can be challenging. Constraints on personnel numbers may restrict the assignment of dedicated maintenance personnel to outside the area of operations. Systems and processes for maintenance must be established and validated to ensure equipment is maintained without degrading the ability to conduct operations. Theater special operations command, theater Army sustainment planners, and the 528th Sustainment Brigade's Army special operations forces liaison elements work together to ensure Army special operations units' sustainment requirements are documented, understood, and resourced appropriately.

6-13. The 528th Sustainment Brigade's Army special operations forces liaison elements are regionally aligned, forward-stationed members of the brigade staff. They provide direct support to theater special operations commands to ensure proper sustainment of Army special operations and sustainment of Army special operations forces supporting joint special operations.

This page intentionally left blank.

Chapter 7
Protection

Protection and preservation has always been an inherent command imperative. All forces, including special operations forces, assist in preserving combat power, populations, partners, resources, and critical infrastructure through protection tasks. Every military activity, from training and predeployment preparation through mission accomplishment, requires the commander to assume responsibility for protecting his force while achieving the objective. Preserving the force includes protecting personnel (combatants and noncombatants), systems, physical assets, and information of the U.S. and multinational military and civilian partners, to include the host nation. The protection warfighting function consists of twelve primary tasks. In addition to the primary protection tasks, commanders and staffs must synchronize, integrate, and organize additional protection capabilities and resources that preserve combat power. The protection tasks that Army special operations unit commanders and their staffs focus on are explained in this chapter.

Note: ADP 3-37 provides additional information on protection.

PERSONNEL RECOVERY

7-1. *Personnel recovery* is the sum of military, diplomatic, and civil efforts to prepare for and execute the recovery and reintegration of isolated personnel (JP 3-50). Personnel recovery is the overarching term for operations that focus on recovering isolated or missing personnel before they become detained or captured or returning human remains for identification and proper burial and honors. As trusted and honorable Army professionals, Army special operations units have a long history of providing support to and conducting unilateral personnel recovery operations. Once in the assigned area of operations, the force can perform its mission unilaterally and with indigenous forces or other governmental departments and agencies to recover isolated, missing, detained, or captured personnel. These units possess the skills, capabilities, and modes of employment to perform personnel recovery missions. The units, in direct support of joint combat search and rescue operations, may be inserted into hostile territory and travel overland to a predetermined rendezvous point to make contact with the evader. Once contact has been made, the recovery force and the evader move to a location within range of friendly assets for extraction.

7-2. Army special operations units are responsible for self-recovery in support of their own operations, consistent with organic capabilities and assigned functions and in accordance with the requirements of the supported commander. These units must make recovery planning an inherent part of every mission and include recovery and emergency exfiltration operations. The vast majority of the recovery planning can be facilitated through the unit's standard operating procedure that is subsequently plugged into the evasion plan of action. Personnel recovery of special operations personnel in a conventional force area of operations may be assisted by the personnel recovery teams organic to Army service component command, corps, or division headquarters and coordinated by the special operations command and control elements in support of those headquarters.

7-3. As a component of the combatant commander's personnel recovery capabilities, Army special operations forces develop two forms of assisted recovery: nonconventional and unconventional. *Nonconventional assisted recovery* is personnel recovery conducted by indigenous/surrogate personnel that are trained, supported, and led by special operations forces, unconventional warfare ground and maritime forces, or other government agencies' personnel that have been specifically trained and directed to establish and operate indigenous or surrogate infrastructures (JP 3-50).

7-4. Nonconventional assisted recovery forces generally deploy into their assigned areas before offensive operations begin and provide the joint force commander with a coordinated personnel recovery capability for as long as the force remains viable. The Joint Personnel Recovery Agency, the proponent for JP 3-50, has approved the following definition of unconventional assisted recovery to be updated in JP 3-50: *personnel recovery operations controlled by resistance force, in conjunction with or facilitate by special operations forces, to return isolated personnel to friendly control.* Using tactics and techniques associated with unconventional warfare and other special operations tasks, unconventional assisted recovery uses resistance-controlled evasion mechanisms to execute recovery operations. Unconventional assisted recovery operations are executed in conjunction with or facilitated by special operations forces to return isolated personnel to friendly control in denied areas where other personnel recovery capabilities are infeasible, inaccessible, or do not exist.

Note: JP 3-05, *Personnel Recovery*, and ATP 3-18.72, (U) *Special Forces Personnel Recovery* (S//NF) provide complete information on the topic.

POPULACE AND RESOURCES CONTROL

7-5. Populace and resources control is conducted in conjunction with and as an integral part of all military operations. Populace and resources control consists of two distinct, yet linked, components: populace control and resources control. These controls are normally the responsibility of indigenous civil governments. Combatant commanders define and enforce these controls during times of civil or military emergency.

Note: FM 3-57, *Civil Affairs Operations*; ATP 3-57.10, *Civil Affairs Support to Populace and Resources Control*; and ATP 3-39.30, *Security and Mobility Support*, provide more information on populace and resources control.

POPULACE CONTROL

7-6. Populace control provides security for the indigenous populace, mobilizes human resources, denies enemy access to the population, and detects and reduces the effectiveness of enemy agents. Populace control measures are a key element in the execution of primary stability tasks in the areas of civil security and civil control.

RESOURCES CONTROL

7-7. Resources control regulates the movement or consumption of material resources, mobilizes materiel resources, and denies materiel resources to the enemy. Resources controls target specific sectors of a nation's material wealth and economy, including natural resources, food and agriculture, immovable property, finances, and cultural and critical infrastructure.

RISK MANAGEMENT

7-8. Commanders use judgment when identifying risks, deciding what risks to accept, and mitigating accepted risks. Commanders accept risks to create opportunities, and they reduce risks by foresight and careful planning. Commanders use risk management to identify and mitigate risks. Risk is a function of the probability of an event occurring and the severity of the event expressed in terms of the degree to which the incident affects combat power or mission capability.

7-9. Safety is a subtask of risk management. It identifies and assesses hazards to the force and makes recommendations on ways to mitigate those hazards. A *hazard* is a condition with the potential to cause injury, illness, or death of personnel; damage to or loss of equipment or property; or mission degradation (JP 3-33). All staffs understand and factor into their analysis how their execution recommendations could adversely affect Soldiers.

7-10. Risk management integration is the primary responsibility of the protection officer or the operations officer. All commands develop and implement a command safety program that incorporates fratricide avoidance, occupational health, risk management, fire prevention and suppression, and accident prevention programs focused on minimizing safety risks.

> *Note:* ATP 5-19, *Risk Management*, provides more information.

OPERATIONS SECURITY

7-11. Although it is not a primary protection task, Army special operations units apply operations security to all operations and activities. Operations security is a process of identifying essential elements of friendly information and subsequently analyzing friendly actions attendant to military operations and other activities to identify those actions that can be observed by adversary intelligence systems; determine indicators that adversary intelligence systems might obtain that could be interpreted or pieced together to derive critical information in time to be useful to adversaries; and select and execute measures that eliminate or reduce to an acceptable level the vulnerabilities of friendly actions to adversary exploitation (JP 3-13.3). All units conduct operations security to preserve essential secrecy. Commanders establish routine operations security measures in unit standard operating procedures. Operations security for these units may require separate reporting and accountability methods.

ARMY HEALTH SYSTEM

7-12. The Army Health System includes health service support (a component of the sustainment warfighting function) and force health protection, which is a component of the protection warfighting function. Force health protection consists of the measures to promote, improve, or conserve the mental and physical well-being of Soldiers, to include those that enable a healthy and fit force, prevent injury and illness, and protect the force from health hazards.

> *Note:* More in-depth Army Health System support information is in FM 4-02, *Army Health System*, and ATP 4-02.43, *Army Health System Support to Army Special Operations Forces*. ATP 3-05.40, *Special Operations Sustainment*, provides more information on unit specific capabilities and requirements.

7-13. Army special operations medical elements, and the command surgeons that oversee them, are an integral part of planning and execution of special operations. In general, the provision of Army health support is challenging as Army special operations units vary in their organic medical capabilities. Assigned medical personnel are relied on to integrate their understanding of the supported plan and the supporting special operations plan, the capabilities of the special operations force, the support and command relationships established by both plans, and their understanding of the operational environment, to include health threats and medical intelligence with all planning and execution efforts.

7-14. Force health protection and special operations medical personnel do not focus solely on operations. The elements of force health protection include enabling health and fitness and preventing of injuries and illness. In addition to unit level internal activities, medical personnel integrate the United States Army Special Operations Command's Tactical Human Optimization Rapid Rehabilitation Reconditioning program as part of a larger United States Special Operations Command mission-specific human performance program. As tactical athletes, Army special operations Soldiers must maintain peak performance in addition to general health and fitness. At the elite tactical athlete level, there are major distinctions in programs designed solely for fitness versus human performance optimization. Tactical Human Optimization Rapid Rehabilitation and Reconditioning increases combat performance and effectiveness, prevents injuries, improves health and longevity, and facilitates rapid return to duty.

Chapter 7

EXPLOSIVE ORDNANCE DISPOSAL

7-15. Explosive ordnance disposal supports, enhances, and enables unit operations, planning, and training across direct and indirect lines of effort. Explosive ordnance disposal competencies specifically provide organic force protection capability, elevate and deepen the level of training assistance to indigenous forces, and reduce risks to missions and to forces during direct action missions, as well as other specialized tasks. Explosive ordnance disposal units provide both general support to Army special operations forces commanders, based on requested theater allocations, as well as direct support with specialized and designated explosive ordnance disposal units to specified units. The increasing proliferation and technological complexity of improvised weapons and other chemical, biological, radiological, and nuclear hazards drives this critical requirement of continued and embedded explosive ordnance disposal support to all Army special operations units in order to maintain required core mission capabilities.

> *Note:* JP 3-42, *Joint Explosive Ordnance Disposal*; ATP 4-32.1, *Explosive Ordnance Disposal (EOD) Group and Battalion Headquarters Operations*; and ATP 4-32.3, *Explosive Ordnance Disposal (EOD) Company, Platoon, and Team Operations*; provide additional information on explosive ordnance disposal.

CHEMICAL, BIOLOGICAL, RADIOLOGICAL, AND NUCLEAR OPERATIONS

7-16. Threat forces are continually attempting to gain possession of and employ chemical, biological, radiological, and nuclear devices in order to disrupt lines of operation. Fear, distrust, and panic among the indigenous population are often the most far-reaching and decisive indirect effects of such an event. The lifecycle of a chemical, biological, radiological, and nuclear event starts with identification of threats, hazards, and vulnerabilities. Once these are identified, counterproliferation operations, such as weapons of mass destruction interdiction, weapons of mass destruction offensive operations, weapons of mass destruction elimination, and chemical, biological, radiological, and nuclear active defense, are conducted to disrupt the enemy's chemical, biological, radiological, and nuclear capabilities. If proliferation prevention fails, passive defense operations are conducted to avoid, protect against, or decontaminate if needed. If overwhelmed, the mission moves to a chemical, biological, radiological, and nuclear consequence management operation, which consists of actions taken to plan, prepare, respond to, and recover from a chemical, biological, radiological, and nuclear incident that requires force and resource allocation beyond passive defense. Throughout the process, diligent and well-integrated chemical, biological, radiological, and nuclear staff sections are necessary to ensure the proper and efficient flow of information, as well as to coordinate higher-level resources to facilitate operations.

7-17. Chemical, biological, radiological, and nuclear operations include the employment of tactical capabilities that counter the entire range of threats and hazards through weapons of mass destruction proliferation prevention, weapons of mass destruction counterforce, and chemical, biological, radiological, and nuclear consequence management activities. These operations support the commander's objectives to counter weapons of mass destruction and operate safely in a chemical, biological, radiological, and nuclear environment.

CHEMICAL, BIOLOGICAL, RADIOLOGICAL, AND NUCLEAR CAPABILITIES

7-18. All of the United States Army Special Command's subordinate commands have U.S. Army Chemical Corps personnel on their staffs. In addition to these staff elements, the 75th Ranger Regiment has decontamination and reconnaissance teams, and each Special Forces Group has a Chemical Decontamination Detachment and a Chemical Reconnaissance Detachment.

7-19. Chemical corps staff personnel have a significant role in the protection warfighting function. They identify chemical, biological, radiological, and nuclear hazards as part of analyzing the operational environment by fusing their analyses with those conducted by the intelligence staff. They identify potential shortfalls during the operations process and facilitate requests for external support requirements. They integrate the use of organic capabilities into special operations during planning and monitor execution of those portions of the operation. The staff ensures interoperability of the unit with forces that either host or support their activities. Common standards for chemical, biological, radiological, and nuclear defense (both training and equipment) are established to maximize cohesion, effectiveness, and efficiency and to prevent or mitigate vulnerabilities in joint force capabilities. This is especially important and challenging during time-constrained contingency operations.

7-20. Because of the regional alignment of Special Forces groups, their organic chemical detachments have unique knowledge and understanding of their assigned area of responsibilities. While the chemical detachments exist to support the group's operations, they may be task-organized to provide support to any special operations task force in order help achieve strategic or operational objectives. Examples of strategic or operational level employment include—

- Determining reconnaissance and surveillance functions in support of special reconnaissance, counter proliferation, and direct action missions.
- Conducting missions to determine the nature, scope, and extent of chemical, biological, radiological, and nuclear hazards activity on specified targets both for prestrike and poststrike activities.
- Assessing, developing, training, organizing or directing foreign chemical, biological, radiological, and nuclear capabilities.
- Providing technical intelligence to other governmental departments and agencies with analysis and presumptive identification of suspect chemical compounds and radioactive isotopes.

Note: ATP 3-05.11, *Special Operations Chemical, Biological, Radiological, and Nuclear Operations*, provides details on Army special operations forces chemical, biological, radiological, and nuclear capabilities and operations.

7-21. Protection of civilians is not a separate task under the protection warfighting function; it is a protection activity that is considered during the planning and execution of all tasks. Army special operations forces often work in close proximity to and/or directly with local populations. This factor requires Soldiers to understand and uphold standards of conduct as well as legal, moral, and ethical obligations. Persons who are neither part of nor associated with an armed force or group nor otherwise engaged in hostilities are categorized as civilians and have protected status under the law of war. Army special operations forces must abide by U.S. policy and comply with the law of war during all armed conflicts—however such conflicts are characterized—as well as during missions conducted across the competition continuum. In order to comply, Soldiers must discriminate between combatants and civilians and take appropriate measures to protect civilians.

7-22. The protection of civilians from deliberate attack is distinct from the legal obligations of U.S. forces to minimize harm to civilians during the conduct of operations. Effective protection of civilians starts before deployment by understanding and acknowledging the complexity of the operational environment, creating rules of engagement and policies that prioritize and account for the protection of civilians, and conducting realistic training that tests Soldiers' decision-making ability. Failure to properly account for and address civilian causalities during tactical operations can create immediate local effects that impact maneuver units and provide opportunities to create global effects through the use of information technologies. Commanders and staff consider the protection of civilians in all activities in order to maintain legitimacy and accomplish the mission.

Note: Department of Defense Law of War Manual and JP 3-0 provide more information on the protection of civilians.

This page intentionally left blank.

Chapter 8
Employment

This chapter provides more context and information on how Army special operations forces support achievement of national objectives while conducting core activities across the range of military operations. The preceding seven chapters provided overarching content on challenges; how U.S. Army Special Operations Command subordinate units cope with those challenges; what special operations are; the principles and tenets used to plan, design, and conduct special operations; the frameworks used by Army special operations forces; and the special operations imperatives. These areas and their relationships with each other were depicted graphically in the Executive Summary. This chapter provides additional focus on support to large-scale combat operations in the context of unified land operations.

OVERVIEW

8-1. The U.S. Army's primary focus is maintaining a ready and lethal force to fulfill its strategic roles. In order to be prepared for large-scale combat operations, the Army focuses its training and priorities on armed conflict against peer and near-peer adversaries. Much of the Army's efforts in the roles of shape, prevent, and consolidate gains contribute to its readiness and ability to fulfill its role of conduct large-scale combat. Within the Army, special operations forces fulfill unique requirements to support government agencies, combatant commanders, allies, and partners across the range of military operations. Army special operations forces support these roles by preparing environments for successful large-scale combat operations and by contributing to the entire joint force's efforts to prevail. Special operations conducted during military engagement, security cooperation, and deterrence are critical for success when the operational environment demands crisis response, limited contingency operations, or armed conflict.

8-2. The range of military operations can provide a foundation to understand the relationships between the strategic uses of force; the Army's strategic roles; individual campaigns, operations and activities; and the Army's special operations force efforts to support achievement of national objectives. These Army special operations force core activities are conducted across the range of military operations as primary or supporting activities, either singly or in combination:

- Civil Affairs operations.
- Countering weapons of mass destruction.
- Counterinsurgency.
- Counterterrorism.
- Direct action.
- Foreign humanitarian assistance.
- Foreign internal defense.
- Military information support operations.
- Preparation of the environment.
- Security force assistance.
- Special reconnaissance.
- Unconventional warfare.
- Hostage rescue and recovery (select Army special operations units only).

8-3. These core activities span across the range of operations, helping to create stable security environments and—when the environment degrades due to crisis or armed conflict—providing a foundation from which positions of relative advantage are obtainable.

8-4. The national objectives supported by Army special operations units are typically related to assessment, shaping, active deterrence, influence, disruption, and threat neutralization. The nature of the strategic environment is such that special operations forces are often engaged in several types of operations simultaneously across a combatant commander's area of responsibility. Figure 8-1 depicts an example of the relationships described in the preceding paragraphs.

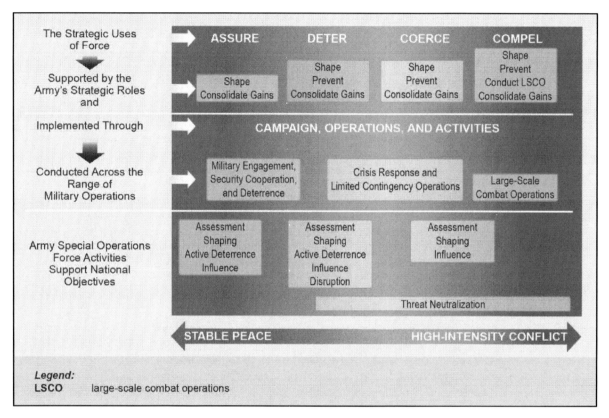

Figure 8-1. Strategic use of force, Army strategic roles, range of military operations, and objectives

8-5. Although Army special operations forces often use unique methods and equipment, the key to success lies with the individual special operations Soldier. In support of a combatant commander's campaign plan, an Army special operations unit may routinely deploy an individual to conduct a specific special operations mission such as providing support to a U.S. Embassy's objectives. The unit may deploy as task-organized teams (such as four individuals), as organic teams, or as a unit. In some cases, the only other Department of Defense presence in the country may be those assigned to the U.S. Embassy. Therefore, it is essential that the plans, orders, and procedures that drive Army special operations forces' employment are clear and direct. Using the approach of mission command assures the commander's intent is understood even for complex operations. Frequent involvement in joint and interagency operations requires an understanding of the U.S. organization for national security and the nature of joint military operations. Special operations forces deployed to a particular theater for various missions (exercises, operations, and support activities) remain under the command authority of the combatant commander or under operational control of the theater special operations command exercised through a subordinate headquarters.

MILITARY ENGAGEMENT, SECURITY COOPERATION, AND DETERRENCE

8-6. The Army's role of shape has led to the use of the phrase "shaping activities." Shaping activities are not defined. A *shaping operation* is an operation at any echelon that creates and preserves conditions for success of the decisive operation through effects on the enemy, other actors, and the terrain (ADP 3-0). Shaping activities is simply a phrase used as a catch all for the broad spectrum of military activities conducted in support of a combatant commander campaign plan. Within the range of military operations, these are binned under military engagement, security cooperation, and deterrence.

8-7. Shaping activities are directly tied to authorities provided in various titles of United States Code (including Titles 10, 22, and 50) and approved programs and are integrated and synchronized with the Department of State, other government agencies, country teams, and ambassadors' plans and objectives. The Department of State and the United States Agency for International Development develop the joint strategy to address regional goals, management, operational considerations, and resources. Each country team develops an individual country plan to address country context, joint mission goals, and coordinated strategies for development, cooperation, security, and diplomatic activities. Combatant commands consider these plans along with functional, regional, and global campaign plans as they develop their campaign plan.

8-8. Army special operations forces shaping activities are typically in support of assessment, influence, and active deterrence related objectives. Shaping activities include those that—

- Establish, influence, maintain, and refine relationships with other nations and foreign and domestic civil authorities.
- Develop information and intelligence.
- Contribute to shared understanding.
- Maintain U.S. influence, access, or interoperability with and to actors or regions.
- Increase ally and partner security force capability and capacity.
- Influence relevant actors.
- Mitigate deteriorating conditions.
- Enhance international legitimacy.
- Gain multinational cooperation.
- Prevent escalation.
- Prepare the environment for future operations.
- Identify, mitigate, or defeat anti-access and/or area denial capabilities.
- Leverage indigenous populations and other human networks.
- Disrupt an adversary's ability to use supporting physical and human networks.
- Disrupt adversarial shadow governments.
- Disrupt adversarial economic, financial, and intelligence infrastructures.
- Disrupt an adversary's ability to make timely and informed decisions.

CRISIS RESPONSE AND LIMITED CONTINGENCY OPERATIONS

8-9. JP 3-0 describes crisis response and limited contingency operations as focused and conducted to achieve a very specific strategic or operational-level objective in an operational area. In the context of a combatant commander campaign plan, crisis response and limited contingency operations can be viewed as branches and sequels within the plan. Typical crisis response and limited contingency operations include noncombatant evacuation operations, peace operations, foreign humanitarian assistance, recovery operations, strikes, raids, homeland defense, and defense support to civilian authorities.

8-10. While a crisis is developing, Army special operations forces may be directed to establish an early forward presence, provide the primary force, or support a larger task force effort. Supported objectives may include assessment, shaping, active deterrence, influence, disruption, and threat neutralization.

8-11. These objectives are achieved through a broad spectrum of activities including those that—
- Prepare the environment for potential follow-on forces or future operations.
- De-escalate or mitigate conditions.
- Initiate military and civilian liaison.
- Assess infrastructure, capabilities, and capacity.
- Establish command and control capabilities.
- Advise friendly forces.
- Provide the combatant commander and subordinate component commanders with an increased understanding of the situation.
- Assure operational access.
- Influence relevant actors.
- Identify, mitigate, or defeat anti-access and/or area denial capabilities.
- Leverage indigenous populations and other human networks.
- Conduct personnel recovery activities.
- Disrupt an adversary's ability to use supporting physical and human networks.
- Disrupt adversarial shadow governments.
- Disrupt adversarial economic, financial, and intelligence infrastructures.
- Disrupt an adversary's ability to make timely and informed decisions.

LARGE-SCALE COMBAT OPERATIONS

8-12. When required to achieve national, strategic objectives or to protect national interests, the national command authority may decide to conduct a major operation involving large-scale combat, placing the United States in a wartime state. In such cases, the general goal is to prevail against the enemy as quickly as possible, to conclude hostilities, and to establish conditions favorable to the United States, its multinational allies, and partners. Large-scale combat operations are usually a blend of traditional and irregular warfare activities relying on offensive, defensive, and stability tasks and activities. Army special operations forces efforts to consolidate gains during the activities that preceded the decision to conduct large-scale combat operations prepare the environment for successful combat operations. The integration of special operations discreet, precise, and scalable principles into unified land operations enables the Army to conduct large-scale combat operations.

8-13. Major campaigns are inherently joint and require robust command and control architectures. The buildup of forces can progress from a battalion-size force (special operations task force) during crisis response or limited contingency operations to a corps-level force (special operations joint task force) during major operations or a campaign. Previously established special operations contingency command and control elements may facilitate the buildup of supporting joint forces. An evolution of force buildup will influence established special operations command and control structures, resulting in a possible operational handover to, and potential integration with, an incoming joint task force. As major operations and campaigns develop, modifying the special operations command and control organization(s) to better enable interdependence and synchronization with the larger U.S. and multinational forces is critical to success. Figure 8-2, page 8-5, provides a graphic depiction of how the command and control structures may change across the range of military operations. Transitions between types of operations, authorities, or phases are opportunities to facilitate planning and operational success. In terms of Army special operations command and control, the following are considerations:
- Make the chain of command clear and concise.
- Create unity of command.
- Include authority to accomplish assigned tasks.
- Clear organizational relationships to facilitate unity of effort.
- Define authorities, permissions, roles, and relationships.
- Ensure every echelon, component, unit, and section understand their authorities, permissions, and relationships.

- Provide supported commands' special operations staff or liaisons with the requisite experience and expertise to plan, conduct, and support operations.
- Ensure liaisons know their roles, responsibilities, authorities, and permissions.
- Integrate special operations forces with conventional, multinational, and host-nation forces early in the planning process.
- Maintain a special operations forces continuity within operational areas and commanders.

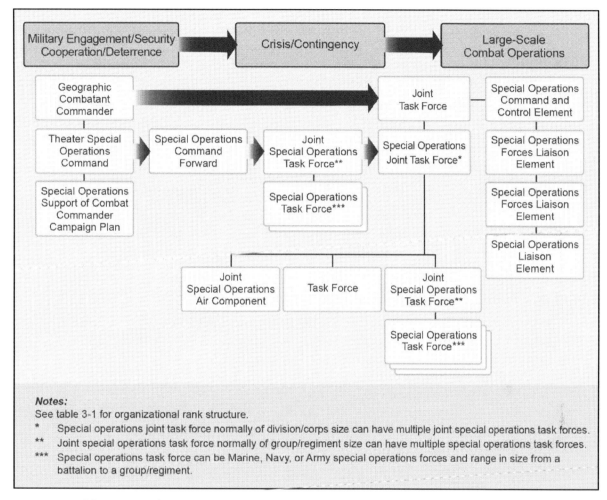

Figure 8-2. Command and control of special operations forces in theater

8-14. In major operations or separate campaigns, the combatant commander will normally establish a joint task force and designate the joint force commander. If designated by the geographic combatant commander, the joint force commander can be a special operations unit commander. When a special operations commander is not designated as the joint force commander, the combatant commander with the theaters' special operations command input will determine requirements for command and control of special operations. This may include directing the commander, theater special operations command, to assume the role of a joint force special operations component commander within the joint force. The role of a joint force special operations component commander and the capability and capacity of a theater special operations command to conduct command and control of special operations forces in large-scale combat operations is finite. Therefore it is likely that Army special operations forces will be required to provide a special operations joint task force. 1st Special Forces Command is charged with the responsibility to establish the core of a special operations joint task force. It is incumbent on the geographic combatant commander to establish appropriate command relationships between special operations units, such as the theater special operations command, joint force special operations component, special operations joint task force, joint special operations task force, and special operations task force.

Chapter 8

8-15. The establishment of a special operations joint task force (as in Operation ENDURING FREEDOM) improves interoperability and interdependence between special operations forces, other joint forces, multinational forces, allies, and partners. The special operations joint task force can plan and coordinate all special operations in the joint operational area, to include employing and sustaining U.S. and multinational special operation forces. The special operations joint task force increases synergies in intelligence, communications, and information sharing, improves manpower efficiency, improves integration of conventional forces and special operations forces, and enhances coordination between all special operations in theater.

8-16. In order to prevail against the enemy, to conclude hostilities, and to establish conditions favorable to the United States and its multinational allies and partners, the special operations joint task force plans, designs, and conducts an integrated supporting special operations plan. The principles, frameworks, tenets, and imperatives described in this publication enable the special operations joint task force commander's subordinate units to conduct special operations—
- In denied areas to leverage indigenous populations and other human networks.
- To open denied areas.
- To facilitate and execute deep operations for joint task force component commanders.
- To provide sensors, combat information, and intelligence from beyond the fire support coordination line (see Chapter 7 for protection considerations).
- To conduct combat identification to inform engagement decisions.
- To conduct direct action to achieve operational and strategic effects.
- To conduct influence operations.
- To attack high-value and high-payoff deep targets.
- To disrupt or defeat threat anti-access efforts.
- To disrupt or defeat threat area denial capabilities.
- To disrupt an adversary's ability to use supporting physical and human networks.
- To disrupt adversary and threat shadow governments.
- To disrupt adversary and threat economic, financial, and intelligence infrastructures.
- To disrupt an adversary's or threat's ability to make timely and informed decisions.
- To counter weapons of mass destruction.
- To conduct personnel recovery activities.
- To consolidate gains achieved by operational success.
- To establish conditions for transitions of control to legitimate civil authorities.
- To facilitate transition to ally and partner nation forces.
- To enable legitimate authority.

SUMMARY

8-17. The Army and the joint force face numerous challenges across a diverse range of operational environments. Army special operations units support unified action and unified land operations to cope with, mitigate, and overcome these challenges. They do this by conducting special operations. Special operations provide strategic, operational, and tactical options for combatant commanders, other joint force commanders, and U.S. Ambassadors. Special operations are guided by three principles (discreet, precise, and scalable) and by two sets of tenets. The Army operations tenets of simultaneity, depth, synchronization, and flexibility and the Army special operations tenets of tempo, preemption, disruption, deception, and disciplined initiative— form the operational art of linking tactical actions to strategic objectives.

Glossary

The glossary lists acronyms and terms with Army, multi-Service, or joint definitions, and other selected terms. Terms for which ADP 3-05 is the proponent publication (the authority) are marked with an asterisk (*). The proponent publication for other terms is listed in parentheses after the definition.

SECTION I – ACRONYMS AND ABBREVIATIONS

ADP	Army doctrine publication
ADRP	Army doctrine reference publication
ATP	Army techniques publication
CJCSI	Chairman of the Joint Chiefs of Staff instruction
DA	Department of the Army
DOD	Department of Defense
DODD	Department of Defense directive
DODI	Department of Defense instruction
FM	field manual
G-2	assistant chief of staff, intelligence
G-3	assistant chief of staff, operations
G-9	assistant chief of staff, Civil Affairs operations
J-2	intelligence directorate of a joint staff
J-6	communications system directorate of a joint staff
JP	joint publication
S-2	intelligence staff officer
STANAG	standardization agreement
TC	training circular
USAJFKSWCS	U.S. Army John F. Kennedy Special Warfare Center and School
USASOC	U.S. Army Special Operations Command
USSOCOM	U.S. Special Operations Command

SECTION II – TERMS

advanced force operations

Operations conducted to refine the location of specific, identified targets and further develop the operational environment for near-term missions. (JP 3-05)

* advanced operations base

A small, temporary base established near or within a joint operations area to command, control, and support special operations training or tactical operations. (ADP 3-05)

* Army special operations aviation

Designated Active Component forces and units organized, trained, and equipped specifically to conduct air mobility, close air support, and other special air operations.

Glossary

Army special operations forces
Those Active and Reserve Component Army forces designated by the Secretary of Defense that are specifically organized, trained, and equipped to conduct and support special operations. (JP 3-05)

* auxiliary
For the purpose of unconventional warfare, the support element of the irregular organization whose organization and operations are clandestine in nature and whose members do not openly indicate their sympathy or involvement with the irregular movement. (ADP 3-05)

characteristic
A feature or quality that marks an organization or function as distinctive or is representative of that organization or function. (ADP 1-01)

Civil Affairs
Designated Active Component and Reserve Component forces and units organized, trained, and equipped specifically to conduct Civil Affairs operations and to support civil-military operations. (JP 3-57)

Civil Affairs operations
Actions planned, coordinated, executed, and assessed to enhance awareness of, and manage the interaction with, the civil component of the operational environment; identify and mitigate underlying causes of instability within civil society; and/or involve the application of functional specialty skills normally the responsibility of civil government. (JP 3-57)

civil-military operations
Activities of a commander performed by designated military forces that establish, maintain, influence, or exploit relations between military forces and indigenous populations and institutions by directly supporting the achievement of objectives relating to the reestablishment or maintenance of stability within a region or host nation. (JP 3-57)

close area
The portion of the commander's area of operations where the majority of subordinate maneuver forces conduct close combat. (ADP 3-0)

* close quarters battle
Sustained combative tactics, techniques, and procedures employed by small, highly trained special operations forces using special purpose weapons, munitions, and demolitions to recover specified personnel, equipment, or material. (ADP 3-05)

commander's intent
A clear and concise expression of the purpose of the operation and the desired military end state that supports mission command, provides focus to the staff, and helps subordinate and supporting commanders act to achieve the commander's desired results without further orders, even when the operation does not unfold as planned. (JP 3-0)

combat search and rescue
The tactics, techniques, and procedures performed by forces to effect the recovery of isolated personnel during combat. (JP 3-50)

consolidation area
The portion of the land commander's area of operations that may be designated to facilitate freedom of action, consolidate gains through decisive action, and set conditions to transition the area of operations to follow on forces or other legitimate authorities. (ADP 3-0)

conventional forces
1. Those forces capable of conducting operations using nonnuclear weapons. 2. Those forces other than designated special operations forces. (JP 3-05)

core competency
An essential and enduring capability that a branch or an organization provides to Army operations. (ADP 1-01)

countering weapons of mass destruction

Efforts against actors of concern to curtail the conceptualization, development, possession, proliferation, use, and effects of weapons of mass destruction, related expertise, materials, technologies, and means of delivery. (JP 3-40)

counterinsurgency

Comprehensive civilian and military efforts designed to simultaneously defeat and contain insurgency and address its root causes. (JP 3-24)

counterproliferation

Those actions to reduce the risks posed by extant weapons of mass destruction to the United States, allies, and partners. (JP 3-40)

counterterrorism

Activities and operations taken to neutralize terrorists and their organizations and networks in order to render them incapable of using violence to instill fear and coerce governments or societies to achieve their goals. (JP 3-26)

deep area

Where the commander sets conditions for future success in close combat. (ADP 3-0)

denied area

(DOD) An area under enemy or unfriendly control in which friendly forces cannot expect to operate successfully within existing operational constraints and force capabilities. (JP 3-05)

direct action

Short-duration strikes and other small-scale offensive actions conducted as a special operation in hostile, denied, or diplomatically sensitive environments and which employ specialized military capabilities to seize, destroy, capture, exploit, recover, or damage designated targets. (JP 3-05)

feasibility assessment

A basic target analysis that provides an initial determination of the viability of a proposed target for special operations forces employment. (JP 3-05)

fires warfighting function

The related tasks and systems that create and converge effects in all domains against the adversary or enemy to enable operations across the range of military operations. (ADP 3-0)

foreign humanitarian assistance

Department of Defense activities conducted outside the United States and its territories to directly relieve or reduce human suffering, disease, hunger, or privation. (JP 3-29)

foreign internal defense

Participation by civilian agencies and military forces of a government or international organizations in any of the programs and activities undertaken by a host nation government to free and protect its society from subversion, lawlessness, insurgency, terrorism, and other threats to its security. (JP 3-22)

guerrilla force

A group of irregular, predominantly indigenous personnel organized along military lines to conduct military and paramilitary operations in enemy-held, hostile, or denied territory. (JP 3-05)

hazard

A condition with the potential to cause injury, illness, or death of personnel; damage to or loss of equipment or property; or mission degradation. (JP 3-33)

insurgency

The organized use of subversion and violence to seize, nullify, or challenge political control of a region. Insurgency can also refer to the group itself. (JP 3-24)

intelligence operations
　　The variety of intelligence and counterintelligence tasks that are carried out by various intelligence organizations and activities within the intelligence process. (JP 2-01)

irregular warfare
　　A violent struggle between state and non-state actors for legitimacy and influence over the relevant population(s). (JP 1)

joint fires element
　　An optional staff element that provides recommendations to the operations directorate to accomplish fires planning and synchronization. (JP 3-60)

joint force special operations component commander
　　The commander within a unified command, subordinate unified command, or joint task force responsible to the establishing commander for recommending the proper employment of assigned, attached, and/or made available for tasking special operations forces and assets; planning and coordinating special operations; or accomplishing such operational missions as may be assigned. (JP 3-0)

joint special operations air component commander
　　The commander within a joint force special operations command responsible for planning and executing joint special operations air activities. (JP 3-05)

joint special operations area
　　An area of land, sea, and airspace assigned by a joint force commander to the commander of a joint special operations force to conduct special operations activities. (JP 3-0)

joint special operations task force
　　A joint task force composed of special operations units from more than one Service, formed to carry out a specific special operation or prosecute special operations in support of a theater campaign or other operations. (JP 3-05)

joint terminal attack controller
　　A qualified (certified) Service member who, from a forward position, directs the action of combat aircraft engaged in close air support and other offensive air operations. (JP 3-09.3)

military engagement
　　The routine contact and interaction between individuals or elements of the Armed Forces of the United States and those of another nation's armed forces, or foreign and domestic civilian authorities or agencies to build trust and confidence, share information, coordinate mutual activities, and maintain influence. (JP 3-0)

military information support operations
　　Planned operations to convey selected information and indicators to foreign audiences to influence their emotions, motives, objective reasoning, and ultimately the behavior of foreign governments, organizations, groups, and individuals in a manner favorable to the originator's objectives. (JP 3-13.2)

nonconventional assisted recovery
　　Personnel recovery conducted by indigenous/surrogate personnel that are trained, supported, and led by special operations forces, unconventional warfare ground and maritime forces, or other government agencies' personnel that have been specifically trained and directed to establish and operate indigenous or surrogate infrastructures. (JP 3-50)

nonproliferation
　　Actions to prevent the acquisition of weapons of mass destruction by dissuading or impeding access to, or distribution of, sensitive technologies, material, and expertise. (JP 3-40)

operational environment
　　A composite of the conditions, circumstances, and influences that affect the employment of capabilities and bear on the decisions of the commander. (JP 3-0)

operational preparation of the environment

The conduct of activities in likely or potential areas of operations to prepare and shape the operational environment. (JP 3-05)

personnel recovery

The sum of military, diplomatic, and civil efforts to prepare for and execute the recovery and reintegration of isolated personnel. (JP 3-50)

preparation of the environment

An umbrella term for operations and activities conducted by selectively trained special operations forces to develop an environment for potential future special operations. (JP 3-05)

Rangers

Rapidly deployable airborne light infantry organized and trained to conduct highly complex joint direct action operations in coordination with or in support of other special operations units of all Services. (JP 3-05)

resistance movement

An organized effort by some portion of the civil population of a country to resist the legally established government or an occupying power and to disrupt civil order and stability. (JP 3-05)

risk management

The process to identify, assess, and control risks and make decisions that balance risk cost with mission benefits. (JP 3-0)

role

The broad and enduring purpose for which the organization or branch is established. (ADP 1-01)

security cooperation

All Department of Defense interactions with foreign security establishments to build security relationships that promote specific United States security interests, develop allied and partner nation military and security capabilities for self-defense and multinational operations, and provide United States forces with peacetime and contingency access to allied and partner nations. (JP 3-20)

security force assistance

The Department of Defense activities that support the development of the capacity and capability of foreign security forces and their supporting institutions. (JP 3-20)

shaping operation

An operation at any echelon that creates and preserves conditions for success of the decisive operation through effects on the enemy, other actors, and the terrain. (ADP 3-0)

sociocultural factors

The social, cultural, and behavioral factors characterizing the relationships and activities of the population of a specific region or operational environment. (JP 2-01.3)

Special Forces

United States Army forces organized, trained, and equipped to conduct special operations with an emphasis on unconventional warfare capabilities. (JP 3-05)

special operations

Operations requiring unique modes of employment, tactical techniques, equipment and training often conducted in hostile, denied, or politically sensitive environments and characterized by one or more of the following: time sensitive, clandestine, low visibility, conducted with and/or through indigenous forces, requiring regional expertise, and/or a high degree of risk. (JP 3-05)

special operations command and control element

A special operations element that is the focal point for the synchronization of special operations forces activities with conventional forces activities. (JP 3-05)

special operations forces
  Those Active and Reserve Component forces of the Services designated by the Secretary of Defense and specifically organized, trained, and equipped to conduct and support special operations. (JP 3-05)

special operations liaison element
  A special operations liaison team provided by the joint force special operations component commander to coordinate, deconflict, and synchronize special operations air, surface, and subsurface operations with conventional air operations. (JP 3-05)

special operations-peculiar
  Equipment, material, supplies, and services required for special operations missions for which there is no Service-common requirement. (JP 3-05)

special operations task force
  A scalable unit, normally of battalion size, in charge of the special operations element, organized around the nucleus of special operations forces and support elements. (JP 3-05)

special reconnaissance
  Reconnaissance and surveillance actions conducted as a special operation in hostile, denied, or diplomatically and/or politically sensitive environments to collect or verify information of strategic or operational significance, employing military capabilities not normally found in conventional forces. (JP 3-05)

* special warfare
  The execution of capabilities that involve a combination of lethal and nonlethal actions taken by a specially trained and educated force that has a deep understanding of cultures and foreign language, proficiency in small-unit tactics, and the ability to build and fight alongside indigenous combat formations in permissive, uncertain, or hostile environments (ADP 3-05).

subversion
  Actions designed to undermine the military, economic, psychological, or political strength or morale of a governing authority. (JP 3-24)

support area
  The portion of the commander's area of operations that is designated to facilitate the positioning, employment, and protection of base sustainment assets required to sustain, enable, and control operations. (ADP 3-0)

* surgical strike
  The execution of capabilities in a precise manner that employ special operations forces in hostile, denied, or politically sensitive environments to seize, destroy, capture, exploit, recover or damage designated targets, or influence threats (ADP 3-05).

terrorism
  The unlawful use of violence or threat of violence, often motivated by religious, political, or other ideological beliefs, to instill fear and coerce governments or societies in pursuit of goals that are usually political. (JP 3-07.2)

theater special operations command
  A subordinate unified command established by a combatant commander to plan, coordinate, conduct, and support joint special operations. (JP 3-05)

time-sensitive target
  A joint force commander-validated target or set of targets requiring immediate response because it is a highly lucrative, fleeting target of opportunity or it poses (or will soon pose) a danger to friendly forces. (JP 3-60)

unconventional warfare

Activities conducted to enable a resistance movement or insurgency to coerce, disrupt, or overthrow a government or occupying power by operating through or with an underground, auxiliary, and guerrilla force in a denied area. (JP 3-05.1)

* underground

A cellular covert element within unconventional warfare that is compartmentalized and conducts covert or clandestine activities in areas normally denied to the auxiliary and the guerrilla force. (ADP 3-05)

weapons of mass destruction

Chemical, biological, radiological, or nuclear weapons capable of a high order of destruction or causing mass casualties, and excluding the means of transporting or propelling the weapon where such means is a separable and divisible part from the weapon. (JP 3-40)

This page intentionally left blank.

References

All websites accessed on 26 August 2019.

REQUIRED PUBLICATIONS

These documents must be available to intended users of this publication.

DOD Dictionary of Military and Associated Terms, as of July 2019.

ADP 1-02, *Terms and Military Symbols*, 14 August 2018.

RELATED PUBLICATIONS

These documents contain relevant supplemental information.

JOINT PUBLICATIONS

Most joint publications are available online at https://www.jcs.mil/Doctrine/.

CJCSI 3110.05F, *Military Information Support Operations Supplement to the Joint Strategic Capabilities Plan*, 7 April 2017. CJCS/JS Limited Access: https://jsportal.sp.pentagon.mil/sites/Matrix/DEL/SitePages/Home.aspx

JP 1, *Doctrine for the Armed Forces of the United States*, 25 March 2013.

JP 2-01, *Joint and National Intelligence Support to Military Operations*, 5 July 2017.

JP 2-01.3, *Joint Intelligence Preparation of the Operational Environment*, 21 May 2014.

JP 3-0, *Joint Operations*, 17 January 2017.

JP 3-03, *Joint Interdiction*, 9 September 2016.

JP 3-05, *Special Operations*, 16 July 2014.

JP 3-05.1, *Unconventional Warfare (FOUO)*, 15 September 2015.

JP 3-07.2, *Antiterrorism*, 14 March 2014.

JP 3-09.3, *Close Air Support*, 10 June 2019.

JP 3-13.2, *Military Information Support Operations*, 21 November 2014.

JP 3-13.3, *Operations Security*, 6 January 2016.

JP 3-20, *Security Cooperation*, 23 May 2017.

JP 3-22, *Foreign Internal Defense*, 17 August 2018.

JP 3-24, *Counterinsurgency*, 25 April 2018.

JP 3-26, *Counterterrorism*, 24 October 2014.

JP 3-29, *Foreign Humanitarian Assistance*, 14 May 2019.

JP 3-30, *Command and Control of Joint Air Operations*, 10 February 2014.

JP 3-33, *Joint Task Force Headquarters*, 31 January 2018.

JP 3-40, *Countering Weapons of Mass Destruction*, 31 October 2014.

JP 3-42, *Joint Explosive Ordnance Disposal*, 9 September 2016.

JP 3-50, *Personnel Recovery*, 2 October 2015.

JP 3-57, *Civil-Military Operations*, 9 July 2018.

JP 3-60, *Joint Targeting*, 28 September 2018.

References

ARMY PUBLICATIONS

Most Army doctrinal publications are available online on the Army Publishing Directorate website (https://armypubs.army.mil).

ADP 1-01, *Doctrine Primer*, 31 July 2019.
ADP 2-0, *Intelligence*, 31 July 2019.
ADP 3-0, *Operations*, 31 July 2019.
ADP 3-19, *Fires*, 31 July 2019.
ADP 3-37, *Protection*, 31 July 2019.
ATP 2-01.3, *Intelligence Preparation of the Battlefield*, 1 March 2019.
ATP 3-05.2, *Foreign Internal Defense*, 19 August 2015.
ATP 3-05.11, *Special Operations Chemical, Biological, Radiological, and Nuclear Operations*, 30 April 2014.
ATP 3-05.40, *Special Operations Sustainment*, 3 May 2013.
ATP 3-07.6, *Protection of Civilians*, 29 October 2015.
ATP 3-18.72, *(U) Special Forces Personnel Recovery (S//NF)*, 13 January 2016.
ATP 3-39.30, *Security and Mobility Support*, 30 October 2014.
ATP 3-57.10, *Civil Affairs Support to Populace and Resources Control*, 6 August 2013.
ATP 4-02.43, *Army Health System Support to Army Special Operations Forces*, 17 December 2015.
ATP 4-32.1, *Explosive Ordnance Disposal (EOD) Group and Battalion Headquarters Operations*, 24 January 2017.
ATP 4-32.3, *Explosive Ordnance Disposal (EOD) Company, Platoon, and Team Operations*, 1 February 2017.
ATP 5-19, *Risk Management*, 14 April 2014.
FM 2-0, *Intelligence*, 6 July 2018.
FM 3-0, *Operations*, 6 October 2017.
FM 3-05, *Army Special Operations*, 9 January 2014.
FM 3-57, *Civil Affairs Operations*, 17 April 2019.
FM 4-02, *Army Health System*, 26 August 2013.
TC 3-05.5, *Special Operations Task Force Planning and Operations Handbook*, 26 September 2017.

DEPARTMENT OF DEFENSE PUBLICATIONS

Most Department of Defense issuances are available online at https://www.esd.whs.mil/DD/.

Department of Defense Law of War Manual, June 2015.
https://dod.defense.gov/Portals/1/Documents/law_war_manual15.pdf
DODD 2060.02, *DoD Countering Weapons of Mass Destruction (WMD) Policy*, 27 January 2017.
DODD 5100.01, *Functions of the Department of Defense and its Major Components*, 21 December 2010.
DODI O-3607.02, *Military Information Support Operations (MISO)*, 13 June 2016.
Department of Defense Strategy for Countering Weapons of Mass Destruction, June 2014.
https://dod.defense.gov/Portals/1/Documents/pubs/DoD_Strategy_for_Countering_Weapons_of_Mass_Destruction_dated_June_2014.pdf

OTHER PUBLICATIONS

STANAG 2523, *Allied Joint Doctrine for Special Operations*, 7 August 2019.
https://nso.nato.int/nso/nsdd/stanagdetails.html?idCover=9045&LA=EN

Allied Joint Publication-3.5, *Allied Joint Doctrine for Special Operations*, 17 December 2013.
https://nso.nato.int/nso/nsdd/apdetails.html?APNo=2589

United States Code, Title 10, *Armed Forces*.

Section 164, *Commanders of Combatant Commands: Assignment; Powers and Duties*, 1 October 1986.
http://uscode.house.gov/view.xhtml?req=(title:10%20section:164%20edition:prelim)%20OR%20(granuleid:USC-prelim-title10-section164)&f=treesort&edition=prelim&num=0&jumpTo=true

Section 167, *Unified Combatant Command for Special Operations Forces*, 18 October 1986.
http://uscode.house.gov/view.xhtml?req=(title:10%20section:167%20edition:prelim)%20OR%20(granuleid:USC-prelim-title10-section167)&f=treesort&edition=prelim&num=0&jumpTo=true

United States Code, Title 22, *Foreign Relations and Intercourse*.
http://uscode.house.gov/view.xhtml?req=granuleid:USC-prelim-title22-front&num=0&edition=prelim

United States Code, Title 50, *War and National Defense*.
http://uscode.house.gov/browse/prelim@title50&edition=prelim

USSOCOM Directive 10-1, *Terms of Reference—Roles, Missions, and Functions of Component Commands*, 9 May 2018.

(*Note:* For organizations outside of USASOC, please send a request to Commander, U.S. Army Special Operations Center of Excellence, USAJFKSWCS, ATTN: AOJK-SWC-DTJ, 3004 Ardennes Street, Stop A, Fort Bragg, NC 28310-9610.)

PRESCRIBED FORMS

This section contains no entries.

REFERENCED FORMS

Unless otherwise indicated, DA forms are available on the Army Publishing Directorate website (https://armypubs.army.mil/).

DA Form 2028, *Recommended Changes to Publications and Blank Forms*.

RECOMMENDED READINGS

Paddock, Alfred H. Jr., *US Army Special Warfare: Its Origins,* National Defense University Press, 1982.

McRaven, William H., *Spec Ops: Case Studies in Special Operations Warfare Theory and Practice,* Presidio Press, 1996,

Index

Entries are by paragraph number.

A

air support, 3-27, 3-47, 4-10, 4-16, 4-17

area assessment, 1-68, 2-39

C

campaign, 1-3, 1-10, 1-19, 1-20, 1-31, 1-34, 1-35, 1-59, 1-60, 2-30, 2-31, 2-43, 2-47, 3-1, 3-14 through 3-16, 3-35, 3-36, 3-40, 4-3, 4-12, 5-15, 6-3, 8-2, 8-5 through 8-7, 8-9, 8-12, 8-13

characteristics, 1-11, 1-13 through 1-16, 2-39

civil-military operations, 1-27, 1-67, 2-4 through 2-6, 2-22, 3-17

clandestine, 1-5, 1-46, 1-68, 2-15, 2-43, 2-47, 2-48

combat search and rescue, 2-19, 4-10, 7-1

command and control, 1-18, 1-56, 1-64, 2-20, 3-3, 3-10, 3-11, 3-15 through 3-18, 3-21, 3-24, 3-30, 3-31, 3-33, 3-34, 4-13 through 4-15, 7-2, 8-10, 8-12, 8-13

core activities, 1-31, 1-38, 2-1, 2-2, 2-28, 3-47, 5-11, 8-2, 8-3

core competencies, 1-9

core principles, 1-46

counterdrug, 1-35

counterinsurgency, 1-27, 2-3, 2-10, 2-12, 8-2

counterproliferation, 1-36, 1-51, 2-8, 2-9, 7-16

counterterrorism, 1-27, 2-9, 2-13 through 2-15, 8-2

covert, 2-15, 2-43, 2-47, 2-48

D

deception, 1-49, 1-55, 1-70, 2-11, 2-28, 3-35, 8-16

F

force multipliers, 1-68

H

humanitarian assistance, 2-21, 2-22

I

imperatives, 1-68, 1-70, 8-15, 8-17

insurgency, 1-33, 1-54, 2-10 through 2-12, 2-23, 2-43 through 2-45, 2-47

J

Joint Chiefs of Staff, 1-6, 1-30, 3-4

joint fires element, 4-10, 4-12, 4-14

joint force special operations component commander, 3-12, 3-14, 3-29, 3-47, 4-13, 8-13

joint intelligence center, 5-1, 5-5 through 5-7, 5-9 through 5-12

joint operations center, 5-10

joint special operations air component commander, 3-27

joint special operations area, 2-41, 3-20

joint task force, 1-17, 1-40, 1-51, 1-57, 1-63, 3-8, 3-12, 3-14, 3-16, 3-17, 3-26, 3-30, 3-32, 3-40, 4-9, 4-12, 5-8, 8-12 through 8-15

L

liaison element, 3-11, 3-29, 4-13, 4-14, 6-7, 6-12, 6-13

liaison teams, 3-31

M

mission criteria, 1-20

mission planning, 1-68

N

national military strategy, 3-1

national security strategy, 1-25, 3-1

nongovernmental organizations, 3-1

O

operation plan, 1-20, 6-7

overt, 1-68, 2-15, 2-47

R

rules of engagement, 1-68, 4-8, 7-22

S

security assistance, 2-24, 3-3, 3-4

special operations command and control element, 3-10, 3-30, 3-31, 3-33, 3-34, 4-13, 4-14

special operations liaison element, 3-29, 4-13, 4-14

subversion, 1-33, 1-54, 2-23, 2-44, 2-45

T

target intelligence packages, 5-7, 5-9

terrorism, 1-33, 2-13, 2-15, 2-23

time-sensitive target, 4-14

W

warfighting functions, 2-28, 4-1, 7-12, 7-19, 7-21

Made in the USA
San Bernardino,
CA